THE
COMPLETE
IDIOT'S
GUIDE® TO

The Sun

by Jay M. Pasachoff, Ph.D.

ALPHA

A member of Penguin Group (USA) Inc.

Copyright © 2003 by Jay M. Pasachoff, Ph.D.

International Standard Book Number: 1-59257-074-7
Library of Congress Catalog Card Number: 2003105470

05   04   03   8   7   6   5   4   3   2   1

Interpretation of the printing code: The rightmost number of the first series of numbers is the year of the book's printing; the rightmost number of the second series of numbers is the number of the book's printing. For example, a printing code of 03-1 shows that the first printing occurred in 2003.

Printed in the United States of America

Most Alpha books are available at special quantity discounts for bulk purchases for sales promotions, premiums, fund-raising, or educational use. Special books, or book excerpts, can also be created to fit specific needs.

For details, write: Special Markets, Alpha Books, 375 Hudson Street, New York, NY 10014.

Publisher: *Marie Butler-Knight*
Product Manager: *Phil Kitchel*
Senior Managing Editor: *Jennifer Chisholm*
Acquisitions Editor: *Mikal Belicove*
Development Editor: *Tom Stevens*
Senior Production Editor: *Katherin Bidwell*
Copy Editor: *Krista Hansing*
Illustrator: *Chris Eliopoulos*
Cover/Book Designer: *Trina Wurst*
Indexer: *Angie Bess*
Layout/Proofreading: *Megan Douglass, Becky Harmon*

Contents at a Glance

Part 1: **What the Sun Looks Like** 1

1 The Sun Shines on Us 3
Sunshine comes from the Sun's surface, but what's on top?
What's being hidden? How do we find out the Sun's secrets?

2 The Active Sun 13
At times the Sun spews out murderous particles almost all
the time, but then it takes a break for a few years. Why?

3 Seeing the Invisible 27
Do you sense the billions of solar particles going through you
every second? Of course not. But you had better find out about
the invisible parts of the Sun and the invisible particles that
it gives off.

4 The Sun Goes Up; the Sun Goes Down 39
Is it too cold for you? Too hot for you? What's this winter and
summer business? And why don't you carry a sundial around
instead of a watch? Are you worried about being late for an
appointment?

5 Our Sun: Looking Good 51
"Blue skies, shining on me," goes the old Irving Berlin song.
But why is the wild blue yonder blue? Why are sunsets red?
And what is in the rainbow?

Part 2: **The Sun Through Time** 61

6 The Sun as a Star 63
Twinkle, twinkle little Sun? No, the Sun is too big for twin-
kling, but it does share a history with billions and trillions of
other stars.

7 The Sun and Civilization 73
We tend to think that we know everything and that there is
nothing new under the Sun. But what did people know about
the Sun hundreds and thousands of years ago?

8 The Birth of the Sun 85
"You must have been a beautiful baby," went the old song. We
see many objects in the sky that are now acting as the Sun did
billions of years ago, when the Sun was but an infant.

9 The Sun at the Center   91
    Our Sun lost its perfection when Galileo got a good look at it
    with a telescope. Copernicus had put our Earth in our solar
    system's center, a sump for undesirable materials. Now our solar
    system is only one of hundreds—which puts us in our place.

10 The Death of the Sun   103
    "There is beauty in the bellow of the blast," sing performers of
    Gilbert and Sullivan operettas. We can now see stars like the
    Sun that are ahead of us in dying. They cast off the most
    beautiful and colorful shells.

Part 3:   Eclipses of the Sun   111

11 Who Stole the Sun?   113
    A member of the ancient Greek Cynics—a school of
    philosophy—told Alexander the Great, who asked if he could
    do him a favor: "Stand aside, you are blocking my light." On
    the contrary, when the Moon gets in our way, it is so beauti-
    ful outside that we don't mind.

12 Saros and Cycles   123
    Do you know where you will be in 18 years, $11\frac{1}{3}$ days?
    Thousands of eclipse buffs do as soon as an eclipse is over.

13 Helium: Only on the Sun   133
    Iron, uranium, hydrogen, coronium. What is wrong with
    that list? Helium may have been found on the Sun, but no
    dice with coronium, in spite of decades of looking for it.

14 To the Ends of the Earth   147
    Solar eclipses haven't been local for the last few years. Scientists
    have hauled equipment all over the world. Tourists go, too,
    since there is incredible beauty alongside the science.

15 To Be in the Moon's Shadow   157
    Some people wouldn't go around the corner to see an eclipse.
    Others flock to the ends of the Earth. There are people of both
    persuasions, though the latter group is right.

16 Venus Tries to Cover Immodestly   169
    Captain Cook went to Tahiti almost 250 years ago to see
    Venus pass in front of the Sun. Hundreds of others went to
    other observing sites. And we will soon get the first chance
    in over 120 years to see such an event.

Part 4:   **The Sun from Mountaintops**   **183**

17   High Above the Clouds   185

Do you want to zap the Hubble Space Telescope? Just point it at the Sun. The computers won't let you do that, but the point is that there are many contrasts between observing the Sun and observing more distant objects.

18   Sunspot, New Mexico, and the House of the Sun   195

Solar astronomers want to observe the Sun where it shines all the time and where it shines clearly. The quest takes them to high and remote mountaintops, places of calm and beauty.

19   Canaries and the Big Dog   205

The Sun shines all around the globe each day, so astronomy is very international. A long list of countries have solar observatories, some of which are great.

20   Ringing Like a Bell   215

When the ground trembles underfoot, is it the subway or an earthquake? The Sun would be trembling if we could stand on it (which we can't); we use the vibrations to find out what is inside.

Part 5:   **The Sun from Space**   **225**

21   Above the Air Is Better   227

What's above the wild blue yonder? What is the atmosphere hiding from us? We have to go there to see. Rockets and spacecraft take us beyond the sky.

22   Sunbeam   237

X-rays of a broken arm give us a different picture than photographs taken in ordinary light. On the Sun, x-rays come from the hottest and most violent regions. We started with rockets, and now we can have spacecraft watching the Sun's x-rays all the time.

23   Yo Ho, SOHO   247

The Solar and Heliospheric Observatory peers at the Sun and catches bits of it as they stream by. SOHO is way out in space, hovering close to the Sun's direction. Its minute-by-minute views in the visible and ultraviolet reveal many mysteries.

24   TRACEing Out the Loops   259

The TRACE satellite's views are so clear that you can imagine yourself up close to the Sun. Though the color in the images is false, golden arches of million-degree gas jostle each other, sometimes lighting up the field of view.

25 Plunging into the Sun   269
*You can't keep us away, even if you are thousands of degrees
hot. It may be tough, but we'll get right up to you, taking
your temperature and checking on you firsthand.*

Part 6:   The Sun-Earth Connection   279

26 Constancy, Thy Name Isn't the Sun   281
*You'd think the solar constant would be constant, wouldn't
you? Well, you'd be wrong. But it was hard to prove it, and
we had to go into space first. Still, a hiccup on the Sun won't
change whether you should wear a sweater tomorrow.*

27 Greenhouses of Salt   293
*Earth is like a flower, in that a greenhouse can help it. But
too much of a good thing isn't helpful anymore. The greenhouse effect on Earth has helped us live, but now we have to
make sure it doesn't run away with itself.*

28 The Forecast Today Is Flares   305
*You don't have to be out in space in your spacesuit to be
affected by space weather. What the Sun gives off can affect
even passengers in high-flying airplanes, not to mention satellites in high orbit. Be careful and watch the forecast!*

Appendixes

A Glossary   315
What things mean, on the ground and in space.

B Solar Observatories   323
*They are spread around the globe, and their latest data are
often online. We provide links.*

C Astronomy Clubs and Solar Interest Groups   329
It's fun for you to join others interested in the Sun. Here's how.

D A Word on Temperature   333
How to convert from one scale to another.

E Selected Readings   335
Do you like this stuff? Here's how to read about more.

Index   339

Contents

Part 1: What the Sun Looks Like   **1**

1 The Sun Shines on Us   **3**

Parts of the Sun 3
Solar Safety 4
The Quiet Sun 5
The Sun's Lower Atmosphere 6
Eclipses and the Sun's Upper Atmosphere 6
Above the Top 8
The Sun as a Star 11

2 The Active Sun   **13**

Dark Spots Are Really Bright 14
Magnets: From Your Refrigerator to Your Sun 15
Sunspot Cycle 17
The Spinning Sun 19
Leading and Trailing 20
Activity Is General and Goes in Cycles *21*
The Spectrum 22
Butterflies 23

3 Seeing the Invisible   **27**

Invisibly Transparent 27
Invisibly Radiating 29
Invisibly Central 30
Invisibly Blocked 33
How and Why the Sun Shines 33
Invisible Neutrinos 35

4 The Sun Goes Up; the Sun Goes Down   **39**

Sunrise, Sunset 40
Solstices and Equinoxes 41
Princess Summerfallwinterspring 44
Figure 8s in the Sky 45
In the Zone 47
Saving Daylight 48
Keeping the Calendar 49

5 Our Sun: Looking Good   51

What Is White? 51
Why Is the Sky Blue? 53
Why Are Sunsets Red? 53
What Is the Green Flash? 54
Rainbows 55
Dogs and Pillars in the Sky 56
Static May Fade "Here Comes the Sun" 57
Rainbows with Lines 57
Planets from Afar 59

Part 2: The Sun Through Time   61

6 The Sun as a Star   63

Presto, the Sun 63
Whither the Sun 66
The Sun Is a Star 69
Spectra Reveal Stars' Secrets 69

7 The Sun and Civilization   73

Stonehenge as an Observatory 73
Where Comes the Sun? 76
Who Found Sunspots? 77
Cathedrals and Sunshine 79
South of the Border 82
Astronomy of "The People" 83

8 The Birth of the Sun   85

Flying Toasters in the Sky 85
Riding Madly Off in All Directions 87
Flickering Starlight 88

9 The Sun at the Center   91

Throw Down Your Ptolemy 91
Galileo on Your Side, Nicky 93
Kepler Controls the Universe 95
Imperfection 98
Dozens and Dozens of Planets 98

10 The Death of the Sun   103
  Red Giants and Yellow Dwarfs 103
  Not a Planet After All 106
  We Won't Blow Up 108
  Our Ultimate Resting Place 109

Part 3: Eclipses of the Sun   111

11 Who Stole the Sun?   113
  Is It a Dragon Eating the Sun? 113
  High Drama in the Sky 117
  Squashed and Flat 119
  The End of Glory 121

12 Saros and Cycles   123
  Almost Too Perfect a Fit 123
    Total and Annular Solar Eclipses 125
    Eclipse Seasons 126
  Over and Over, but Not Quite the Same 127
  Eclipses and World Travelers 130

13 Helium: Only on the Sun   133
  Historical Eclipses in Science, Literature,
    and Politics 133
    October 16, 1876 B.C.E. 134
    June 5, 1302 B.C.E. 134
    585 B.C.E. 135
    Lunar Eclipse of February 29, 1504 135
    April 22, 1715 135
    October 27, 1780 136
    May 15, 1836 136
    July 16, 1806 137
    1879 with No Eclipse, and June 21, 528 137
    1885 138
  Only on the Sun 138
    Helium on the Sun 139
    Coronium in the Corona? 140
  A Laboratory in the Sky 141
    Magneto-hydro-dynamics 142
    Forbidden Lines 144
    Solar Neutrinos 145

The Complete Idiot's Guide to the Sun

14 To the Ends of the Earth   147
    As Long as Can Be 148
    Rooting for Extreme Cold 150
    Caribbean Sun 150
    Eclipses in Europe 151
    Recent American Eclipses 152
        1984 Annular Eclipse in the South 152
        1994 Annular Eclipse Crossing Diagonally 153
        Partial Eclipses 154
    Eclipses and Elephants 154
        2001 in Southern Africa 155
        2002 in Southern Africa 155
    Outback Eclipse 155

15 To Be in the Moon's Shadow   157
    What Does the Future Bring? 157
        Only in Antarctica 158
        No Land—Maybe Just a Little 159
        Turkey? 159
        Way Up North, Plus 159
        As Good as It Gets 160
        Home Again 160
        Once More, with Passion 161
    The Dregs 161
        October 14, 2004 162
        April 8, 2005 162
    Eclipse Viewing and Photography 162
        Still Photography with Film 162
        Still Photography with a Digital Camera 165
        Eclipse Videography 165

16 Venus Tries to Cover Immodestly   169
    A Blot on the Face of the Sun 169
        Good Things Come in Pairs 172
        The Black-Drop Effect 173
        Lengthy Expeditions 174
        American Expeditions 175
        Not Venus's Atmosphere 176

x   The Complete Idiot's Guide to the Sun

14 To the Ends of the Earth   147

Mercury to the Rescue 177
Our Time Has Come 179
   The Transit of 2004 ....*179*
   The Transit of 2012 ....*180*

Part 4: **The Sun from Mountaintops**   **183**

17 **High Above the Clouds**   **185**
The Sun Shines in the Daytime 185
The Sun Isn't Up at Night 187
   Higher Telescopes *189*
   Cutting Down on Turbulence *190*
Let It All Come Through 191

18 **Sunspot, New Mexico, and the House of the Sun**   **195**
Southwest Slant 195
   The Largest in the World *196*
   Hold Still *196*
   Out of Air *197*
   Maui's House *198*
   Corona Without an Eclipse *199*
   Coronagraph Tricks *200*
California Dreaming 201
Synoptic SOLIS 202

19 **Canaries and the Big Dog**   **205**
Gone to the Dogs 205
Indian Idyl 208
The Other Alps 209
Where, Oh Where, Will My Telescope Go? 210

20 **Ringing Like a Bell**   **215**
Images, Plus or Minus 215
   Devastation on the Richter Scale *217*
   In and Out *218*
What's Behind? 220
'Round the World in Six Stops 222

Part 5:   The Sun from Space   **225**

21 Above the Air Is Better   **227**

The Air Is Blocking Your View 227
The Sun and Radar 228
High Above the Atmosphere 228
People Aloft 229
Solar to the Max 232
Spacelab 233
The Voyage of Ulysses 234

22 Sunbeam   **237**

Imaging X-Rays 237
Solar Rockets 239
Yohkoh's Decade 241
Launching Yohkoh 241
Using Yohkoh 241
The Death of Yohkoh 242
Yohkoh Science 242
Exploring Small 243
Go GOES 245

23 Yo Ho, SOHO   **247**

Europe and America Conjoined 247
SOHO's Coronal Science 249
LASCO Blocks the Center 249
Going to Extremes 252
Fast vs. Slow 254
Whizzing Ions 254
Diagnosis 254
Blowing in the Wind 255
SOHO Peers In 255
Swinging Away 255
Trapped 255
Not a Virgin 256
ERNE but No Bert 257

24 TRACEing Out the Loops   **259**

You'll Think You're Up Close 260
Ultra and Beyond 262

The Sun Is Loopy 263
Fire Burn and Cauldron Bubble 265
No Rolling Stones 267

25 Plunging into the Sun   **269**
B Follows A 269
   B's Visual Views 270
   B's X-Ray Views 271
   B's Ultraviolet 271
STEREO Views 271
Activity Over Time 273
Solar Probe 274
The Longer Term 275
   Solar Polar Imager 275
   Not a Dodge in Your Future 275
   Solar Orbiter 276
Can We Do It? 277

Part 6: The Sun-Earth Connection   **279**

26 Constancy, Thy Name Isn't the Sun   **281**
The Solar Parameter? 282
At All Altitudes 282
Is the Sun Going Away? 283
Monitoring Our Lifeline 286
Chiaroscuro 287
Is the Sun Round? 289
The Sun and Weather 290

27 Greenhouses of Salt   **293**
Energy In and Out 294
   The Terrestrial Greenhouse 294
   Greenhouse Gases 295
Not Your Ordinary Greenhouse 295
   We Are Changing Our Atmosphere 298
   What Can We Do? 299
Hole in the Sky 301

28 The Forecast Today Is Flares   **305**

How's the Space Up There? 305
Twisted Magnetism *307*
Flares and Other Eruptions *309*
Ejecting Matter 310
Space Weather 311

Appendixes

A Glossary   **315**

B Solar Observatories   **323**

C Astronomy Clubs and Solar Interest Groups   **329**

D A Word on Temperature   **333**

E Selected Readings   **335**

Index   **339**

Foreword

Humans have scrutinized and wondered about the Sun since the dawn of civilization. There are many reasons for this intense, continuing interest. The Sun sustains life on Earth; it controls our environment and impacts our technological civilization. Understanding and predicting the influences of the Sun on the Earth's climate and on space weather in the near-Earth environment are major challenges for solar astronomy. Solar and stellar evolution continues to influence the evolution of life on Earth and planetary systems in general. The Sun is the nearest and most readily studied example of many physical processes that form the foundations of our current understanding of the universe. The Sun is a unique, astrophysical-scale plasma physics laboratory. It is close enough for laboratory-accurate measurements to be made. The Sun presents us with many important mysteries and unexplored domains that challenge science. *The Complete Idiot's Guide to the Sun* captures the essence of these phenomena and much more in a manner that is easily understood. It forms a basis from which the reader can easily launch further studies of the Sun and astronomy.

It was not without some angst that I undertook writing a foreword to Jay Pasachoff's book—*The Complete Idiot's Guide to the Sun*. Jay is such a prolific writer of astronomy texts that a foreword almost seems superfluous. Jay is one of the few, or perhaps only, authors who could pen this book, which is certainly in line with his recent Education Prize from the American Astronomical Society, "for his willingness to go into educational nooks where no astronomer has gone before." He has brought together a wide range of astronomical research and knowledge—using the Sun as a catalyst—showing how we apply physical insight gained from the Sun in our interpretation of the cosmos, and how we use our knowledge gleaned from other parts of the universe to unravel solar mysteries. Jay's many years of teaching all aspects of astronomy give him an in-depth perspective on the Sun's role as a cornerstone of astronomy. He has done a wonderful job using this perspective to produce a very readable and enlightening book.

Very appropriately, the first two figures of the book include a solar eclipse. No one has devoted more effort than Jay to viewing solar eclipses and bringing solar experiments into the path of totality or to expressing enthusiasm to others for these impressive events. I recall reading a tongue-in-cheek article back in my graduate student days that used the existence of total solar eclipses to prove there is a divinity, who by arranging for total solar eclipses revealed to us much of the workings of the universe. Within our solar system, it is only from the Earth's surface, where the relative angular size of a moon and the Sun are nearly identical, that one can stand on a planetary surface and view a perfect blocking of the bright solar disk to see chromosphere and corona.

Jay covers many topics you might find only in advanced college textbooks—such as the reason the Sun apparently has a sharp edge—but explains them in terms of everyday experience. He provides one of the best and easiest-to-follow explanations of opacity I have read. The book often takes you by surprise; you think this is going to be too hard to explain, and suddenly, very clearly, you find yourself in the heart of nucleosynthesis or helioseismology and it all seems easy and logical. A few pages later, you find yourself understanding the solar neutrino puzzle and its solution.

Jay quickly has the reader doing astrophysics in his or her head (no computers), for example computing the wavelength dependence of the Rayleigh scattering of light, leading to an understanding of why the sky is blue and explaining the mystery of the green flash. The book is dotted with interesting historical notes, letting us follow the progress of solar physics and astronomy in general from ancient times, through the Middle Ages, through the Renaissance, and into modern times.

The Complete Idiot's Guide to the Sun is timely. Recent high-energy spectroscopic images of the Sun's outer atmosphere made from satellites and breakthroughs in resolving fundamental physical processes by using adaptive optics at large ground-based telescopes have placed us at the brink of understanding fundamental processes on the Sun. Although there is still a lot we don't know about the Sun, Jay's book does an excellent job of describing what we do know and pointing out where understanding is still lacking.

Even though the book is called an idiot's guide, Jay manages to squeeze in much of astronomy, proving that the Sun not only provides for life on Earth, but it also provides us with a continuing source of knowledge. *The Complete Idiot's Guide to the Sun* is a fun read, which I enjoyed immensely.

—Dr. Stephen L. Keil

Dr. Stephen L. Keil is director of the National Solar Observatory and Principal Investigator for the Advanced Technology Solar Telescope (ATST), a planned 4-meter aperture solar telescope that will revolutionize our understanding of magnetic process on the Sun. Prior to joining the NSO in 1999, Dr. Keil led the Air Force's Solar Environmental Disturbances task and was the program manager for the Solar Mass Ejection Imager, which is now in orbit around the Earth. He is a 1969 graduate of the University of California and earned his doctorate in Physics and Astronomy at Boston University in 1975. He has over 60 publications in scientific literature.

Introduction

Do you enjoy going outside on a sunny day? Do you prefer the Sun to the gloom? I certainly do, for both my personal life and my professional life. And when I am outside, I am confident that I know something about what is going on over my head. The Sun is shining, and I know why.

As you read this book, you will find out why, too. But the Sun is more than just a nuclear furnace safely set 93 million miles away from us. Energy travels from its center to about 70 percent of the way out, in just the way a space heater sends out energy. But then the last 30 percent or so of the energy is carried out by a type of boiling. And how do we know these facts? The ideas came from theory, but the details have been pinned down in recent years by studying ripples and oscillations on the Sun's surface.

I like the Sun's exterior more than its interior. My own work especially studies the normally invisible gas that we can detect only at eclipses or from spacecraft. While we get aesthetic pleasure from admiring the beautiful shapes of the gas, we also get intellectual pleasure by making progress in understanding its temperature and motions. How does it get to be millions of degrees hot, for example? It has been my academic pleasure on more than one occasion to take a dozen students and a ton and a half of equipment, most recently to Australia, to try to find out why and how at a total solar eclipse.

I have observed the Sun in a variety of ways—using telescopes on eclipse expeditions, on mountaintops, and in space. I hope you will enjoy reading about how we study the Sun as much as I enjoy carrying out the studies and writing about my own work and that of others.

In this book, you will see these notes:

The Solar Scoop
Things you should know about the Sun and other objects we discuss.

Sun Safety
Cautionary remarks, given that the Sun is so bright that you shouldn't ever stare at it—except during the seconds or minutes of the total phase of a total solar eclipse.

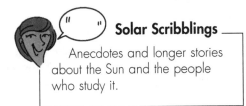

Solar Scribblings

Anecdotes and longer stories about the Sun and the people who study it.

Sun Words

The terms we use in talking about our nearest star.

Fun Sun Facts

Particular items of interest, often out of the ordinary train of thought.

Acknowledgments

The author would like to acknowledge the assistance of Leon Golub (Harvard-Smithsonian Center for Astrophysics), Peter Foukal, John Harvey (National Solar Observatory), John Leibacher (Gong Project, National Solar Observatory), and Spencer Weart (American Institute of Physics) with scientific matters; Deborah Pasachoff and Naomi Pasachoff with proofreading; Alan Title and Zoe Frank (Lockheed Martin Advanced Technology Center/Solar and Astrophysics Laboratory), Nigel Sharp (National Optical Astronomy Observatories), David McKenzie (Montana State University), and many others acknowledged herein for supplying photographs; Daniel B. Seaton (Harvard-Smithsonian Center for Astrophysics) and Utsav Kc, Sarah Croft, and Galen Thorp (Williams College) for gathering photographs; and Linda Crowe for editorial consulting. I am pleased to acknowledge the important role in my life of Naomi Pasachoff, Deborah Pasachoff, Eloise Pasachoff, and, newly, Tom Glaisyer.

Special Thanks to the Technical Reviewer

The Complete Idiot's Guide to the Sun was reviewed by an expert who double-checked the accuracy of what you'll learn here, to help us ensure that this book gives you everything you need to know about the Sun. Special thanks are extended to Dr. Leon Golub.

Trademarks

All terms mentioned in this book that are known to be or are suspected of being trademarks or service marks have been appropriately capitalized. Alpha Books and Penguin Group (USA) Inc. cannot attest to the accuracy of this information. Use of a term in this book should not be regarded as affecting the validity of any trademark or service mark.

Part 1

What the Sun Looks Like

If you can't tell your belly button from your neck when somebody asks you about them, you can't talk about your body. In this first part of the book, we discuss what the Sun is like and the basic terms that we use to describe its structure. We talk about the everyday Sun that warms us with its sunshine and about the rare, beautiful glimpses that we get of the Sun's normally invisible layers at eclipses. We find out how magnetism shapes the sunspots. And we find out how we see inside the Sun, delving beneath its opaque surface. We discuss how the Sun moves in the sky, and even why the sky is blue. For your delight, we even talk about the green flash and rainbows.

Chapter 1

The Sun Shines on Us

In This Chapter

- ◆ The Sun compared with Earth
- ◆ How to watch the Sun safely
- ◆ What the quiet Sun is like daily
- ◆ What is above the Sun's visible surface

Whether or not you can find Orion, Andromeda, or other constellations, you can surely find the Sun in the sky. It's up in the daytime, so you don't even have to stay up late to see it. The ease of finding the Sun is but one of many reasons why the Sun is the most interesting astronomical object. Unlike the stars, the Sun noticeably changes from day to day, so there is always something going on up there. Furthermore, these changes on the Sun sometimes lead to explosions that end up affecting us on Earth. So although stars and galaxies are part of our universe, the Sun is an important part of our backyard.

Parts of the Sun

We on Earth have our feet on the ground but our heads in the air. The Sun's "ground" isn't as sturdy as our Earth's because the Sun is a giant ball of gas through and through. But the Sun also has what we call a surface and an atmosphere. The transparent part is the atmosphere, for the Sun as for the Earth. What we call the "surface" of the Sun is like the Earth's surface in that we can't see through it. And both the Earth and the Sun have interiors that are hidden from our view. In both cases, though, studying the surface in detail can reveal facts about the interior.

The Sun's surface is about a million miles across. Since the Earth is about 100 times smaller across, more than a million Earths could fit inside the Sun. From a distant star, looking back at the Sun, we would see giant planets like Jupiter and Saturn long before we could find a relatively tiny one like our Earth.

The Sun, with sunspots, photographed on an ordinary day through a special filter that cuts out 99.999 percent of the sunlight.

(National Solar Observatory/ AURA/NSF)

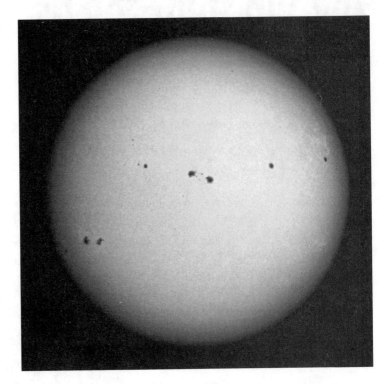

Solar Safety

You shouldn't stay out too long in the Sun, or you'll get sunburned and even start on your way to skin cancer. But looking at the Sun directly with your eyes is much more immediately hazardous. The Sun is so bright that on a normal sunny day, your eye blinks very quickly if the Sun comes into your field of view, and you turn your gaze away. So you wouldn't normally stare at the Sun. If you forced yourself to do so, you could cook your eyeball and cause permanent damage to your vision.

If you tried to look at the Sun with binoculars or a telescope, the Sun's rays would be concentrated into your eye. You could get a permanent blind spot immediately. So think about the Sun and look at its pictures in this book, in other books, or on the World Wide Web, but don't try to look at it directly unless you know how to do so safely.

Sunglasses or other ordinary filters, including even dark filters for cameras, don't make it safe for your eyes to stare at the Sun. Only special Sun filters, which pass only about one part in a million of the incoming sunlight, are safe to look through. (We'll say more about this in Chapter 5.)

Sun Safety

Never stare at the everyday Sun. It is so bright that you could damage your eyes temporarily or even permanently.

It is good to be excited about the Sun, but don't be so excited that you run out and hurt your eyes.

The only time you can ever look at the Sun directly is during the total part of a total solar eclipse. But no total solar eclipse will be visible from the United States or Canada until the year 2017. So unless you travel to see an eclipse, the rule is: No staring at the Sun.

The Quiet Sun

The *quiet Sun* is the way the Sun looks on a steady basis, which is basically how the Sun shines on a daily basis. In the next chapter, we will talk about the active Sun, the way the Sun changes from day to day and in a cycle over the years.

Sun Words

The **quiet Sun** is the everyday, nearly unchanging background Sun.

We humans, and almost all living creatures on Earth, get our light and heat from the Sun. Without sunlight, the Earth would be a cold, dark ball. The amount of energy we get for a given area on Earth (square meter, square inch, whatever) is called the *solar constant*. Wouldn't you know it? Space research has shown in the last decade or so that the solar constant isn't constant after all, but it is steady enough for us to have a standard value that goes with the quiet Sun.

Sun Words

The **solar constant** is the amount of energy that reaches a square meter (or other defined unit of area) of the top of the Earth's atmosphere each second.

The Solar Scoop

Do you have solar panels on your roof? Black pipes to absorb sunlight? If you have solar heating, you depend on more than just the solar constant. Scientists measure the solar constant as the amount of energy coming from the Sun each second in a square meter at the top of the Earth's atmosphere. Before that energy reaches your roof, it can be hidden or diffused by dust, clouds, or other impediments.

The Sun's Lower Atmosphere

The sunlight that we get comes from the top layer of the ball of gas that we call the Sun. Since the Sun is a *sphere*, which means a round ball in the Greek language that gave so many words to English, and the Greek word for "light" is *phot*, the layer that gives us sunlight is called the *photosphere*. Thus, *photosphere* is just the sphere from which the light comes. Really, it is only a shell and not a whole sphere, since all the light comes from the top layer. That layer is only a few hundred miles (alternatively, a few hundred kilometers) thick, which makes it a very thin shell surrounding an object a million miles (1.4 million kilometers) across.

What is the photosphere? It is a layer of gas that is heated enough to shine. Like all the Sun, the gas is about 90 percent hydrogen, about 9 percent helium, and less than 1 percent a mixture of the rest of the chemical elements. Heat trillions of tons of this gas up to 10,000°F (5,500°C—see Appendix D for more on temperatures), and it shines brightly. The photosphere gives off most of its energy in the middle part of the spectrum that we can see with our eyes, but it also gives off a lesser amount of every other kind of energy.

Sun Words

The **photosphere** is the surface that we see in visible light.

Eclipses and the Sun's Upper Atmosphere

Sometimes, about once every 18 months in a 100-mile-wide path across some parts of the Earth, the Moon hides the photosphere entirely. Only when this everyday surface is completely hidden do we have a *total eclipse* of the Sun. Total eclipses have long helped astronomers find out what the otherwise hidden parts of the Sun are like. The basic parts of the Sun were revealed and classified about 150 years ago, in the latter half of the nineteenth century.

Sun Words

A **total eclipse** occurs when the Moon entirely hides the Sun.

The white solar corona silhouetting the Moon, which appears dark, during the total solar eclipse of 2002.

(Jay M. Pasachoff and the Williams College Eclipse Expedition)

When the Sun's everyday surface is just barely covered in a total eclipse, a thin reddish band circles the Sun. Since the Greek word for color is *chroma*, and the reddish color seems to jump out at you after the whitish sunlight disappears, that part of the Sun is known as the *chromosphere*. It remains visible only for a few seconds at an eclipse, no matter how far you have traveled or how hard you try.

After a few seconds of a total eclipse, the chromosphere is covered up, and a halo of white light becomes visible around the Sun. It has really been there all along but has just been too faint to see. During a total eclipse, the sky becomes dark as night, or, at least, as dark as twilight. The darker sky allows the halo, known as the *corona* from the Latin word for *crown*, to be seen with the naked eye.

Sun Words

The **chromosphere** is the colorful shell around the Sun that pops into view at the beginning and the end of a total eclipse. The **corona** is the pearly white crown of light visible around the Sun at a total eclipse.

The parts of the solar atmosphere, the outer layers of the Sun.

Chromosphere

Corona

Photosphere

Though the everyday surface of the Sun, the photosphere, is 5,800 kelvins (K), which may seem hot to you, the chromosphere can be twice as hot. And the corona can be 50 times hotter, millions of kelvins. A typical temperature for the corona is 2 million kelvins, which is about 4 million degrees Fahrenheit. One of the things that my team does at total solar eclipses is try to figure out why the temperature becomes so high in the corona.

Edmond Halley, for whom Halley's comet is named, described the corona as "pearly white"—that is, as milky white as a pearl of the quality used for jewelry. That description has stuck.

Sun Words

A **coronagraph** is a type of telescope that makes artificial eclipses.

Above the Top

The solar corona that we see at eclipses extends a million miles or so into space, which is only about 1 percent of the distance between the Sun and the Earth. But a device is on board the Solar and Heliospheric Observatory (SOHO) spacecraft that makes an artificial eclipse of the Sun. Such a device is called a *coronagraph*. The coronagraphs working on SOHO are part of an experiment called the Large Angle Spectroscopic Coronagraph, or LASCO. They were built and are controlled by scientists at the U.S. Naval Research Laboratory in Washington, D.C.

Coronagraphs can't produce an image as good as those made at natural eclipses, but they can do some things that natural eclipses can't. In particular, the coronagraphs on SOHO have to hide the inner part of the solar corona, although they give hourly images of the corona outside that. The innermost coronagraph, so-called C1, no longer works. But C2 shows the Sun's corona from 1.5 to 6 solar radii, and C3 shows the Sun's corona from 3 solar radii to 32 solar radii. Thirty-two solar radii is 16 solar diameters, or about 16 million miles. So, using LASCO, we can see the corona as it extends about $^{16}/_{93}$ or $^1/_6$ of the way between the Sun and Earth.

The outer solar corona, imaged with the outermost of the three coronagraphs on SOHO. A disk blocks the center of the Sun and the inner corona; the size of the solar photosphere is outlined as a white circle.

(Naval Research Laboratory)

Still, the outer layer of the Sun, the corona, extends still farther. It is expanding, making what is called the *solar wind*, particles blowing outward from the Sun. We will say more about the solar wind when we discuss space weather. The solar wind was discovered by

Sun Words

The **solar wind** is the corona expanding into space.

analyzing the tails of comets and noticing that they moved about as though buffeted by a wind. Now we can study it directly, with spacecraft placed out in space to measure the solar particles as they go by.

A comet's tail is shaped by the solar wind. Here we see Comet Hale-Bopp, a comet that was bright to the naked eye in 1995. Comets this bright may appear only every decade or two.

(Jay M. Pasachoff)

The outer solar atmosphere goes even farther out. In fact, it goes past all the planets. Where it meets the magnetic field of planets, the planets are protected by a "bow shock" (pronounced to rhyme with "bow-wow," not "bow and arrow"), just as a boat moving through water has a shock wave in the water around its bow and trailing behind. But other than that, the Sun's influence extends billions of miles out into space. The region in which the Sun's influence dominates is called the *heliosphere*.

Sun Words

The **heliosphere** is the zone in which the Sun has more influence than interstellar space. The **heliopause** is the location where the interstellar space becomes more important than the solar influence; it marks the end of the heliosphere.

Two NASA *Voyager* spacecraft, which visited Jupiter, Saturn, Uranus, and Neptune, are still in touch with Earth, sending back messages about the conditions of the outer parts of the heliosphere. (Contact has been lost with two, earlier, *Pioneer* spacecraft, in 1995 and 2003, respectively.) They are beyond the orbits of Neptune and Pluto but haven't found the edge of the heliosphere yet. That edge, where the influence of interstellar space matches the influence of the Sun, is called the *heliopause*. Scientists expect that the spacecraft will pass entirely through the heliopause in the next few years. We have some indications that radio signals that we sent out have been bounced back by the heliopause.

The Earth is enveloped in a safety cocoon, in which our magnetic field interacts with the magnetic field of the heliosphere, protecting us on the Earth's surface from the solar particles.

The Solar Scoop

For a discussion of the *Pioneer* and *Voyager* missions through the outer solar system to the heliopause and beyond, see solarsystem.nasa.gov/missions/jup_missns/jup-p11.html and links from it.

The Sun as a Star

The photosphere, chromosphere, and corona are the standard parts of the Sun. They are all we can see on a regular basis. In some sense, they are the basic Sun. We are particularly interested in studying them not only for themselves, but also because the Sun is a typical star in many ways. Billions of stars are hotter than the Sun, and billions of stars are cooler. Billions of stars are intrinsically brighter than the Sun, and billions are intrinsically fainter. And whatever we see on our Sun no doubt occurs on uncountable billions of other stars.

The Sun is a wonderful laboratory for scientists. It has nuclear fusion, but at a definitely safe distance of 93 million miles (150 million kilometers). It has layers, which are too narrow to be seen on distant stars but which must be on them as well. As we shall see, it even gives off particles that we detect on Earth. Similar particles must be hitting planets around other stars.

Our Sun is the dominant body in what we call our solar system. The solar system consists of nine or so major planets, tens of thousands of minor planets, and lots of smaller or less substantial bodies, such as meteorites and comets.

Perhaps the major discovery of the last decade is that our solar system is not the only system of planets around stars. More than 100 planets have been discovered orbiting other stars. We call them "extrasolar planets," or *exoplanets*, using the Latin sense of *extra-* as "outside" or "beyond." For most of these systems, which we can still call "solar systems," even though they are around other suns than our own, we know of only one giant planet, often much more massive than our own solar system's Jupiter. Our discoveries continue, now finding less massive planets,

Sun Words

Exoplanets are planets found around stars other than our Sun.

although they are still approximately the mass of Jupiter or Saturn. It may be a decade or two before our methods are sensitive enough to detect planets as puny as Earth, but we are sure they are there.

Anyway, because we have discovered planets and, in a few cases already, systems of several planets around other stars, we see that here, too, our Sun is typical. Our discoveries about the Sun must be multiplied by billions to show how much we are learning about all the stars in our universe.

The Least You Need to Know

- ◆ Never stare at the Sun, except during the total phase of a total solar eclipse.
- ◆ When we look at or image the Sun normally, we see its photosphere.
- ◆ The Sun's upper atmosphere includes the chromosphere and the corona.
- ◆ The corona expands into space as the solar wind, filling the heliosphere.
- ◆ More than 100 planets have been found orbiting other stars.

The Active Sun

In This Chapter

- ◆ Sunspots look dark only because they are surrounded by brighter stuff

- ◆ Sunspots are cooler than the surrounding stuff

- ◆ Sunspots show where the Sun's magnetic field is strongest

- ◆ Sunspots vary with a cycle about 11 years long

- ◆ Many other things on the Sun also vary with the same cycle as the sunspots

Although you should never stare at the Sun, over thousands of years there have occasionally been times when there is just the right amount of haze in the Earth's atmosphere that the Sun hasn't appeared too bright. Sunspots were discovered at such times even before they were first seen with telescopes in 1610, by Galileo and others.

Anyway, on the surface of the Sun—now studied with special filters or with electronic cameras—dark spots appear. These spots appear tiny on the surface, although they are really often larger than Earth. They have long been known as sunspots, a name applied in the time of Galileo about 400 years ago.

For about 150 years, we have known that the number of sunspots on the Sun increases and decreases—we say that they wax and wane—with a period of about 11 years. That period is known as the *sunspot cycle*. When we are at the minimum of the sunspot cycle—called sunspot minimum—no spots may be visible on the Sun for days at a time. But when we are at the maximum of the sunspot cycle—called "sunspot maximum"—a hundred or more sunspots could be visible at a time. And of course, there are probably just as many sunspots hiding on the Sun's far side.

Sun Words

The **sunspot cycle** is the 11-year up-and-down in the number of sunspots.

Dark Spots Are Really Bright

Photos of sunspots show that they have dark centers that are surrounded by alternating bits of light and dark material. The darkest part of a sunspot is its umbra, named from the Latin word for "shadow." The somewhat lighter region around it is its penumbra, named from the Latin words for "almost shadow."

A sunspot has a dark center known as the umbra, and lighter surrounding parts that are known as the penumbra.

(T. Rimmele/NSO/AURA/NSF)

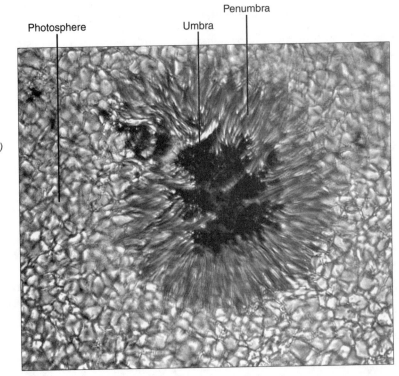

Photosphere   Umbra   Penumbra

Sunspots come in a variety of sizes, but even ordinary ones on the Sun are much larger than the Earth. If you could somehow take one of those sunspots off the Sun and hold it out in space all by itself, it would shine as brightly as the full Moon. It merely *looks* dark because it is surrounded by even brighter gas.

Sunspots are about 2,000 kelvins cooler than the photosphere that surrounds them. So they are about 4,000 kelvins instead of about 6,000 kelvins. But that makes them a lot dimmer because the amount of energy that anything gives off goes down with the temperature raised to the fourth power. So although the temperature in the photosphere is $6,000 \div 4,000 = 6 \div 4 = 1.5$ times hotter, it is $(6,000 \div 4,000)^4 = 1.5^4 = 1.5^2 \times 1.5^2 = 2.25 \times 2.25 = 5$ times brighter. (We won't do anything in this book that requires a calculator. We'll use only math that you can do in your head. That will be accurate enough for us.)

Magnets: From Your Refrigerator to Your Sun

What's a magnet? It is something that pulls certain materials toward it or pushes them away. It uses magnetism, which is one of the basic forces of nature. Scientists have known for more than a hundred years that magnetism and electricity are kindred forces, different aspects of a unified force known as electromagnetism. We use electricity and magnetism on Earth every day—every motor in your house or your car relies on both, for example.

Magnets attract only certain materials or materials in certain conditions. Iron, for example, is attracted to magnets. Just hold a magnet up to an iron nail. The nail will even jump across a bit of air to get to the magnet, if the magnet is strong enough.

To see what the attraction by a magnet looks like, get some iron filings. I used to think that they were something special, but you can really get them (unless you want to write away to, or order on the Internet from, a scientific supply company) merely by filing away a nail with a regular file. Put the magnet under a piece of paper and drop the filings on top of the paper. Sometimes tapping the paper lightly with your finger helps the iron filings find their way to overcoming friction that holds them to the paper.

You will see a pattern of lines on the page. These are "magnetic lines of force." They go from one part of the magnet, called a north pole, to another part of the magnet, called a south pole. Some complicated magnets may have more than two poles, but never do we find only one pole. You never find just a north pole or just a south pole.

Magnetic lines of force around a bar magnet, produced by dropping iron filings on a piece of paper that is placed over the magnet.

(Jay M. Pasachoff)

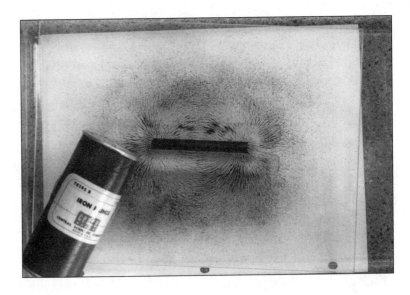

On the Sun, moving hot gas sets up *magnetic fields.* In particular, in sunspots, a very strong magnetic field appears. That magnetic field holds gas in place, since the gas can move only along the direction of the magnetic field lines and not across them. In this way, the magnetic field acts as a giant hand, tamping down motions. It cools the gas. And when the gas is cooler, it is darker.

Sun Words _____

A **magnetic field** shows the direction and strength of magnetism.

We can actually see the shape of magnetic field lines on the Sun, as we could see the lines from a refrigerator magnet if we used it to attract iron filings. A wonderful NASA spacecraft called Transition Region and Coronal Explorer (TRACE) can see detail on the Sun so fine that the magnetic lines of force show. We are actually seeing the glowing gas that is following along the shapes of the magnetic lines of force.

Fun Sun Facts

You need a special filter or a special telescope to look at spots on the Sun. A little telescope device called a Sunspotter is available for about $300. It folds—bounces—the incoming beam of sunlight back and forth so that the device is only about 18 inches high. You can point it toward the Sun to give you a sunspot image projected on a piece of paper. Setup time is only about 10 seconds. See www.starlab.com/ltiss.html.

How strong is the magnetic field in a sunspot? The overall, average magnetic field on the Sun's surface is about 1 gauss, and the magnetic field of a toy magnet is about 100 gauss. But in a sunspot, the magnetic field is 3,000 gauss, much higher. This high magnetic field keeps the sunspots cool and, therefore, dark. (In the international system of units now in scientific use, tesla, which are 10,000 times larger than gauss, are mandated. Since these units are named after people, their symbols are capital letters—T and G, respectively. So perhaps we should be saying that a sunspot, which we said was 3,000 G, is 0.3 T, whereas an ordinary toy magnet is about $\frac{1}{10,000}$ T. Note that the symbols are capital letters, though the units gauss and tesla are written with lowercase letters, because both Gauss and Tesla were people.)

Fun Sun Facts

You can use binoculars or a telescope to project an image of the Sun onto a piece of paper or cardboard or even onto the ground. But you must be very careful not to look up through the optical device at the Sun. If you are using binoculars, extended use could cook the optics inside. In any case, block off the big end of half of the binoculars with cardboard carefully taped on with heavy tape. Then hold the device (telescope or binoculars) about 18 inches in front of the cardboard, with the big end of the binoculars toward the Sun. *Remember: Don't look through the binoculars at the Sun.* Instead, use the binoculars to project the solar image onto a piece of paper. Focus or move the device toward and away from the Sun until you see a sharp solar image. Don't stare at the image for more than a few seconds at a time, since it will be very bright even though it is merely a projection.

Sunspot Cycle

When Galileo discovered sunspots, some of his contemporaries found them, too. In fact, priority for discovering sunspots was part of Galileo's fight that led to his house arrest. Anyway, finding a sunspot was a big thing in the seventeenth century. It often led to publication of a scientific paper, with the *Philosophical Transactions of the Royal Society* of London as a prestigious source. (It is fun that *Phil Trans* is now available online back to volume 1 in 1665, albeit by subscription.)

It wasn't until the mid-nineteenth century, about 150 years ago, that an amateur astronomer noticed that there were actually sometimes a lot of sunspots on the Sun and that the number of sunspots rose and fell with a detectable period. The peaks of the numbers of sunspots occurred about every 11 years, and the times when there were almost no sunspots also differed by about 11 years. The period isn't a constant 11 years—sometimes the interval is only about 8 years, and sometimes it is 12 years or more, but 11 years has proved to be a good average.

The Sun, with sunspots, from a book by Galileo in 1613.

(Jay M. Pasachoff)

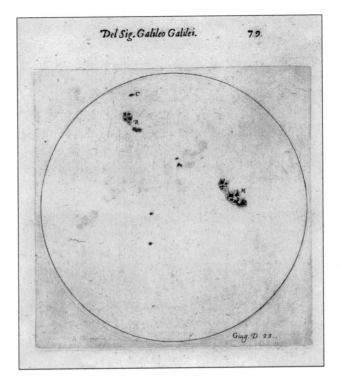

This 11-year period is known as the *sunspot cycle*. It isn't one of those minor effects that scientists sometimes find in which you have to use careful or statistical methods to tease a scientific conclusion out of messy data. No, the sunspot cycle just knocks your eyes out.

For historical reasons, we usually plot not the actual number of sunspots on the Sun, but rather the "sunspot number." This value, also called the Wolf sunspot number, after its nineteenth-century inventor, takes note of the fact that sunspots don't usually appear in isolation on the Sun. Rather, they appear in groups. The sunspot number gives 10 times more importance to groups than it does to individual spots.

Let's say there are four sunspots on the Sun. If they could be scattered around the Sun evenly (this could never happen for magnetic reasons, as we shall soon discuss), there would be four groups, each containing one spot. Thus, the sunspot number would be 44: 4 groups × 10, plus 4 spots. If, as is more likely, there were two regions each with two spots, the sunspot number would be 24: 2 groups × 10, plus 4 spots. And if the four spots were in the same group, the sunspot number would be 14. Still, when the sunspot number is high, there are a lot of sunspots, and also a lot of groups of sunspots, on the Sun.

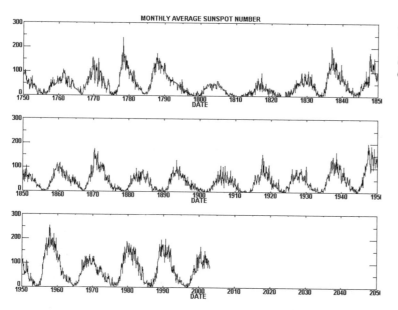

The sunspot cycle.

(NASA's Marshall Space Flight Center)

Some peaks of the sunspot cycle are higher than other peaks. We don't know why. Some say that there is a longer cycle that is superimposed on the shorter cycle. But there is no proof of that. In fact, a couple of years before the most recent sunspot maximum in 2001–2002, there was a session at a scientific meeting at which three papers were given. The first said that the peak would be higher than average, the second said that the peak would be lower than average, and the third said that the peak would be average. (It wound up about average, though lower than the two previous peaks.)

The Spinning Sun

One of my friends was making an IMAX movie about the Sun—and you should see it in an IMAX theater, if you can, or buy the videotape or DVD, if you can't. The movie, called *SolarMax*, contains fabulous footage taken with telescopes on the ground and in space. But my favorite part of the movie is the set of images taken from a book I own: *Macchie Solare* (that is, *Sunspots*), written by Galileo and published in Rome in 1613. My friend aligned and animated a series of photos of full-page drawings that Galileo had published. The movie shows the sunspots moving as the Sun rotates. (My name was only slightly misspelled in the movie's credits.)

So the Sun rotates on its axis—that is, spins—about once a month. Actually, the Sun isn't a solid body like Earth, so all of it doesn't have to rotate in the same amount of time. And that is what happens. Things on the Sun's equator rotate in about 25 days, while things halfway north or south rotate about 2 days a month slower. Things near the pole take more than 28 days to go around. Two and a half days out of 25 is about 10 percent, which may seem like a slight difference. But for every 10 rotations of the equator, sunspots at higher latitudes go around 1 time fewer. (This differential motion also occurs as you go deeper into the Sun.) So if you were to start with a magnetic line of force stretched from pole to pole, the equatorial part of it would gradually but inexorably go faster around, and would eventually lap the higher latitudes. After some years, it would go around a few extra times. The lines of force get wound up in this way.

Now find a rubber band. Hold one part in your left hand and the opposite part in your right hand. Then twist your hands in opposite directions. You will see the rubber band start to kink, and a piece will push up higher. This is the same effect that makes sunspots. Magnetic lines of force at or under the Sun's surface kink as a result of the Sun's equator rotating faster than its poles. When a kink in the magnetic lines of force pushes up through the surface, we see one polarity where it pierces the surface from below and the opposite polarity when it goes back down.

This rotation of the Sun that is different at different latitudes is called differential rotation.

Leading and Trailing

Looking at a map of solar magnetism, you may not notice a particular effect at first. But after a while, it hits you: The black regions are to the right of the white regions in the top hemisphere, while they are to the left in the bottom hemisphere. Because the Sun is rotating with the side facing us going from left to right, we say that the sunspot in the direction of rotation is "leading" and the behind sunspot is "trailing."

Sunspots always appear in groups that contain both polarities. That is because magnetic lines of force always have to start on one polarity and end on the opposite polarity. Thus, though the lines of force don't show on this overhead view, they extend from the bright regions to the nearby dark regions. Sometimes bright and dark are well separated, while other times they are jumbled. We will see later on that the jumbling can lead to explosions known as solar flares.

Solar Scribblings

The magnetic field of sunspots was studied and mapped out about 100 years ago by George Ellery Hale. Hale started as a solar scientist but went on to be the greatest telescope builder for astronomy in general, with his work on the 100-inch telescope on Mount Wilson and the 200-inch telescope on Palomar Mountain.

Notice also that in the northern hemisphere, white polarity leads black. But in the southern hemisphere, it is the other way around. Which polarity leads is always different between the two hemispheres. This effect is known as Hale's Polarity Law, after George Ellery Hale. After each 11-year cycle, the polarity that leads switches in each hemisphere. Thus for 11 years, white (north, say), polarity leads in the northern hemisphere, and then black polarity leads for the next 11 years. As a result, the real sunspot cycle is 22 years long. Only then are the polarities back the way they were before.

Fun Sun Facts

How do astronomers measure magnetism on the Sun, which is 93 million miles away? They use an effect discovered 100 years ago by the Dutch scientist Pieter Zeeman. This Zeeman effect causes a difference in some special parts of the Sun's spectrum, which we will be discussing later. This difference depends on how strong the magnetic field is. Using the Zeeman effect, scientists make maps of the magnetism over the whole Sun. The image shows one polarity as dark and the other as bright. (Astronomers don't always even note which is north and which is south; only the fact that they are opposite counts.)

Activity Is General and Goes in Cycles

The sunspot cycle was historically the first cycle to be found on the Sun. But now we know that a whole host of things on the Sun vary with the same period. All these things result from the Sun's magnetic field. We say that rather than just a sunspot cycle, we have a *solar-activity cycle*.

If you use a special kind of solar filter that passes radiation only from hydrogen, the regions around sunspots are brighter than average. They have long been given the name plage (pronounced *plahje*, after the French word for "beach"). This kind of filter costs $500 or more each—and, of course, professional ones cost a lot more than that. Anyway, following the plages shows not only the Sun's rotation, but also, over years, the solar-activity cycle.

If you can get above Earth's atmosphere, which we can with spacecraft, you can study parts of the radiation from the Sun that doesn't come through Earth's atmosphere.

Sun Words

The **solar-activity cycle** is the sunspot cycle matched in other solar phenomena.

The Spectrum

Our eyes are tuned to what we call visible light, which extends from red through orange to yellow, to green, to blue, to indigo, and to violet. However, in 1800, the astronomer Williams Herschel discovered the *infrared*.

Infrared is a kind of electromagnetic radiation, like light. It works just like light, but infrared waves are more stretched out than light waves. We say that it has longer wavelengths. At still longer wavelengths, we reach radio waves.

Going the other way from the solar spectrum, we go beyond the violet end to get ultraviolet light, which has shorter wavelengths than violet, which in turn, has shorter wavelengths than blue. At still shorter wavelengths, we get extreme ultraviolet (XUV or EUV to astronomers). And at even shorter wavelengths, we get first x-rays and then gamma rays. Much of these radiations don't come through the Earth's atmosphere. In particular, gamma rays, x-rays, and ultraviolet are blocked. Most of the infrared is blocked. But visible light, some of the infrared wavelengths, and radio waves pass right through the Earth's atmosphere, and we can study them on the ground.

Starting in the late 1940s, using captured V-2 rockets, *x-ray* telescopes have been launched above the atmosphere to study the Sun. (One of the pioneers in this work shared in the 2002 Nobel Prize in physics.) Once x-ray telescopes got advanced enough to see details on the Sun, scientists learned that the regions around sunspots are bright in x-rays. The number and position of these regions varies with the solar-activity cycle. Current satellites aloft, like the SOHO and TRACE, make images in the extreme ultraviolet. These images, too, show bright regions of solar activity. The number and brightness of these images wax and wane with the solar-activity cycle.

> **Fun Sun Facts**
>
> Think of the friendly fellow ROY G. BIV, the initials of the color names, to remember the colors. In 1800, the astronomer William Herschel (who had earlier discovered the planet Uranus) put a thermometer beyond the red on a spectrum. The temperature went up. Herschel had discovered *infrared* (that is, beyond the red, from the Latin prefix *infra-*).

> **Sun Words**
>
> **X-rays** are like light, only hundreds or thousands of times shorter in wavelength.

When the magnetic field gets all twisted up in active regions, energy is stored up. Sometimes something triggers an explosion, which releases all the stored energy in seconds. The result is a *solar flare*. Solar flares eject matter at high speed as well as emitting x-rays and *gamma rays*, all of which reach Earth in minutes to days. They are a major source of space weather. The number and strength of solar flares is a major indicator of the solar-activity cycle. We have the strongest flares and the greatest number of them near the peak of the solar-activity cycle, and few flares—and weak ones, at that—at the solar-activity minimum.

A much less explosive phenomenon is also often seen at the edge of the Sun. These are *prominences*. They glow especially reddish, like the chromosphere, and can also be seen at solar eclipses. Even without eclipses, special telescopes can follow the chromosphere and prominences. Sometimes prominences are seen to erupt, though they do so over hours instead of the seconds of flares. Very often, you will see a picture in the newspaper or on the web labeled "flare," while it was really a prominence. Anyway, prominences also follow the solar-activity cycle.

Sun Words _____

A **solar flare** is a powerful, sudden eruption on the Sun. **Gamma rays** are like light but have even shorter wavelengths than x-rays.

Sun Words _____

Prominences are structures held in space above the Sun by the Sun's magnetic field and seen when they are on the edge of the Sun. **Filaments** are what prominences look like when you see them from above.

When you look right at the disk of the Sun rather than at its edge, using hydrogen light that shows prominences well at the Sun's edge, you see dark streaks curving across the surface across plages and between sunspots. These are *filaments*, prominences seen from above. Though prominences look bright because we are seeing them projected against the dark sky, filaments look dark because we are seeing them projected against the bright solar surface. Sometimes when a flare erupts halfway across the Sun, a few minutes later you see a filament wink off and on. It has been temporarily disrupted by an abrupt wave of pressure given off by the flare. Such waves are known as shock waves.

Butterflies

As we discussed, the sunspot cycle is caused by kinks in magnetic lines of force. As time goes on, the kinks go closer to the Sun's equator. Over the 11-year cycle, we can see the sunspots start at latitudes of about 50° north and south, and wind up at latitudes close to 10° north and south. (They seldom reach exactly the equator.)

About a hundred years ago, the British astronomer E. Walter Maunder plotted not only the sunspot number but also the latitude of the spots. For each vertical line on his graph, the latitudes where there are sunspots on a given day or in a given week or month are graphed. The horizontal axis shows the date.

Sun Words

The **butterfly diagram** is the graph that shows butterfly shapes when the latitude of sunspots is graphed over time.

Looking at Maunder's graph, or updated versions of it, shows how the sunspots start at high latitudes and wind up at low ones. As they reach low latitudes, the next sunspot cycle starts at high latitudes. For reasons that are obvious while looking at the graph, it is known as Maunder's *butterfly diagram.*

The butterfly diagram, showing the change over the sunspot cycle of the latitude at which sunspots appear.

(NASA's Marshall Space Flight Center)

DAILY SUNSPOT AREA AVERAGED OVER INDIVIDUAL SOLAR ROTATIONS

SUNSPOT AREA IN EQUAL AREA LATITUDE STRIPS (% OF STRIP AREA)   ■ > 0.0% ■ > 0.1% ▢ > 1.0%

90N
30N
EQ
30S
90S
    1880 1890 1900 1910 1920 1930 1940 1950 1960 1970 1980 1990 2000
                                         DATE

AVERAGE DAILY SUNSPOT AREA (% OF VISIBLE HEMISPHERE)

0.5
0.4
0.3
0.2
0.1
0.0
    1880 1890 1900 1910 1920 1930 1940 1950 1960 1970 1980 1990 2000
                                         DATE

http://science.msfc.nasa.gov/ssl/pad/solar/images/bfly.gif   NASA/NSSTC/HATHAWAY 2003/02

The butterfly diagram is kept up to date in various places, notably on the web by David Hathaway of the Marshall Space Flight Center in Huntsville, Alabama. See science.msfc.nasa.gov/ssl/pad/solar/sunspots.htm for the latest graphs of the sunspot cycle and of the butterfly diagram. Hathaway also includes predictions for the rest of the sunspot cycle. His diagram shows that we should reach sunspot minimum in 2007. Sunspot maximum should then follow five or six years later.

Fun Sun Facts

Helios was the Sun god in Greek mythology. So, *heliosphere* merely means "the sphere of the Sun."

The Least You Need to Know

◆ Each sunspot has a dark umbra surrounded by a lighter penumbra.

◆ Sunspots are about 2,000 kelvins cooler than the surrounding photosphere.

◆ The sun rotates on its axis in about 25 days at the equator and slower toward the poles.

◆ The sun's differential rotation causes sunspots.

◆ The polarity of sunspot that leads in each hemisphere switches every 11 years.

◆ As the sunspot cycle wears on, sunspots appear closer to the equator.

Seeing the Invisible

In This Chapter

- ◆ Why we don't see right through the Sun
- ◆ How parts of the spectrum other than visible light help us learn about the Sun
- ◆ How solar seismology tells us about the Sun's inside
- ◆ Why the Sun shines
- ◆ How neutrinos escape from the Sun's center and tell us about it and about themselves

When we look up at the Sun, we see right through the chromosphere and corona. That's why we see the photosphere, even though those layers are between it and us. They are just invisible to us because they are too thin and insubstantial.

Invisibly Transparent

Imagine that you are in a smoky room (thankfully, something that is becoming rarer). The smoke may be so dense that you can't even see people on the far side of the room. But what if someone walks over to you from there? Halfway over, he or she may become visible. So the smoke has optical thickness but isn't entirely opaque.

Astronomers use the idea of optical thickness to explain how much they see through a gas. When a gas has optical thickness of zero, then it has no optical thickness and is completely transparent. Read on to see what it means for optical thickness to be greater than zero.

A couple important and interesting numbers come up all the time in mathematics. One is π (pi), which stands for the ratio of the circumference of a circle to its diameter. It is approximately equal to 3.14159265358979323 … and cannot ever be expressed exactly as a decimal. In mathematicians' terms, that makes it an irrational number.

The other important irrational number is e, which is approximately equal to 2.71828459045 …, and which for our purposes, we can think of as "about 3."

For an astronomer, a gas has an optical thickness of 1 when the intensity of the light that gets through is reduced by a factor of the special number e, which is about 2.7. A gas has an optical thickness of 2 when the intensity of the light that gets through is reduced by a factor of $e \times e$, which is about 8. So if gas between you and something has optical thickness of 2, only about $1/8$, or about 10 percent, of the light gets through. Even less gets through optical thickness of 3, and for higher optical thicknesses, it is hardly worth considering the tiny amount of light that gets through. The total *opacity* of the gas is increasing as we look through more of it.

> **Fun Sun Facts**
>
> As of 2003, more than a trillion places of pi have been calculated, confirming that no pattern in the digits exists, at least up to that point. The calculations are carried out in large part to test computer methods.

> **Sun Words**
>
> A gas's **opacity** is a measure of how opaque—not transparent—it is.

This calculation of optical thickness means that we see into the Sun until the gas between it and us gets to about optical thickness of 2 or so, and we don't see anything of optical thickness less than about $1/2$. In fact, for many purposes, we can consider the photosphere as occurring at optical thickness of $2/3$.

But optical thickness of $1/2$ up to 2 occurs in the photosphere only when we are considering all the light from the Sun together, which we call *white light*. If we look through a special filter that shows only certain colors that have higher optical thickness, then the optical thickness piles up to be noticeable before we reach the photosphere, looking down. In particular, if we use a red filter that passes only the red light from hydrogen, there is so much hydrogen on the Sun that we reach optical thickness 2 while we are still in the chromosphere. So the chromosphere becomes visible to us, even with a telescope on Earth, and we have found a way of overcoming its general transparency.

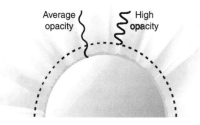

The depth at which we stop seeing into the Sun depends on the color of the light.

Once we have used this filter, we can use ordinary cameras—film or electronic—to take images. The light that comes through is in the visible, and film and electronics are sensitive to it. We merely leave out all the other colors of light that hide this color.

Fun Sun Facts

If we look through some gas and its optical thickness is less than about a half, the gas looks transparent to us. But if we look through some gas whose optical thickness is greater than about 2, it looks opaque. So the question why the Sun has a sharp edge is related to the following: What is the angle near the edge of the Sun, when looking from Earth, between a line of sight that has an optical thickness of $1/2$ and a line of sight that has an optical thickness of 2?

The answer to that question is that the angle is very small—smaller than we can see with our unaided eye. So the Sun goes from transparent to opaque over an imperceptibly small angle for us, and the edge of the Sun looks sharp.

Invisibly Radiating

Light, the term we use for optical radiation, has a few special colors that allow us to see the chromosphere. These colors correspond to hydrogen gas in the red or blue and to calcium gas in the violet, just on the edge of ultraviolet. If we look through filters that pass only these specific colors, the Sun's gas can be more opaque. With this higher opacity, we stop seeing at higher heights.

If we look at certain wavelengths that the SOHO or TRACE spacecraft use, the photosphere does not emit strongly and we see coronal gas silhouetted against a dark background, even though its total opacity is not high. Again, we have used special techniques to reveal what was otherwise invisible.

An image taken by the SOHO spacecraft's Extreme-ultraviolet Imaging Telescope through a filter that passes only a narrow set of wavelengths in the ultraviolet. The image shows the solar corona; the background photosphere doesn't radiate much at this wavelength.

(EIT Team, NASA's Goddard Space Flight Center)

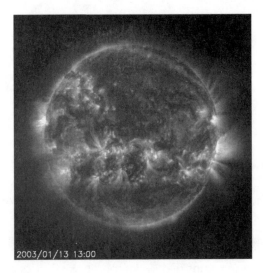

2003/01/13 13:00

Invisibly Central

Below the photosphere, things are invisible for a more fundamental reason: Our view of them is blocked by the photosphere itself. We can't see down to the center of the Sun, and certainly not down to its core, in ordinary light. Even x-rays or radio waves don't go through the photosphere, so they aren't of any help in studying the core.

Solar astronomers now study the center of the Sun by using a technique used by geologists and other geoscientists on Earth. They use a type of earthquake wave. On Earth, geologists can't peer through the Earth's crust, but they use devices called seismometers to measure waves that are caused by earthquakes and other disturbances. As those waves pass through parts of the Earth, they are changed and distorted. By measuring the earthquake waves that reach a seismometer—or, better, a network of seismometers spaced around the Earth—geoscientists can tell what structures the Earth had between the earthquake and their measuring instruments.

Similarly, solar astronomers can measure waves on the surface of the Sun. They do this by taking pictures of the Sun every few seconds. By comparing the brightness and velocity of individual bits of sunlight—individual pixels—they can tell how each bit of the Sun changes over time. They then use some standard mathematical methods to discover what periods of waves are involved.

Calculations show that large areas of the Sun are moving up and down, and that the period of the oscillation depends on the size of the regions.

(National Solar Observatory, NOAO/AURA/NSF)

Since similar methods on Earth are called seismology, the work on the Sun is called "solar seismology," or *helioseismology*. Over the last two decades, helioseismology has grown to be a major part of solar astronomy. It has found that the Sun vibrates with various periods—that is, it rings like a bell.

One problem with helioseismology is that the Sun sets at night. To find out what waves of long periods—hours or days—exist on the Sun, you have to observe for long periods consecutively. Having the Sun set in the middle ruins your stream of data. The problem has been solved in two ways.

Sun Words _____

Helioseismology is the science of finding out about the inside of the Sun by studying oscillations of the Sun's surface.

The first way is just to find a way to keep the Sun from setting—figuratively, at least. The solution started decades ago with observations made from the U.S. base at Thule, Greenland, where at certain times of year, the Sun doesn't set. Fighting clouds and weather, consecutive runs of 80 hours and later longer were achieved.

A second way to keep the Sun from setting is to distribute telescopes around the Earth. After all, it used to be said that the Sun never set on the British empire. Now the Sun never sets on the identical telescopes of the Global Oscillation Network Group. The group's initials spell GONG, which is suitable for something that is finding vibrations with a variety of periods. (We will say more about it in Chapter 20.)

Fun Sun Facts

Sometimes professionals who study the Sun are called solar astronomers, but even more often they are called solar physicists. The two terms come from the origin of the study at the turn of the twentieth century. People who studied the Sun tended to come from the physics population, perhaps because they were studying the Sun's spectrum or using techniques like infrared. Astronomers of the time tended to study where the stars were, and they worked at night instead of in the daytime. The term "solar physicist" stuck. To this day, the main professional journal in the field is called *Solar Physics*.

The GONG group is an international consortium based at the National Solar Observatory in Tucson, Arizona. It has identical telescopes spaced around the world, in places such as Udaipur, India and Learmouth, Australia. It has now operated for several years and has continuous chains of observations that are months long. Version GONG++ upgraded the original telescopes to take data more often and on a finer scale of spatial resolution—that is, able to see finer details.

A third way of keeping the Sun from setting is to go out into space. The SOHO spacecraft sits a million miles above Earth in the general direction of the Sun. We say that it is in a "halo orbit," in that it moves in a small halo-like circle around a point between the Earth and Sun where the gravity balances. If it were directly between the Sun and us, we couldn't get data back from it because the background static from the Sun would overwhelm the spacecraft's radio.

SOHO carries three instruments that do helioseismology. From its position in space, the Sun is always visible, so the instruments make continuous sets of observations. Except for brief intervals when they malfunction, they work consecutively for months or years.

One of the results of all this helioseismology is that we understand the center of the Sun very well. We have learned how fast the Sun rotates at different distances from its center. These data have shown us how the Sun's differential rotation extends way down from the surface. We have also learned that the Sun has an outer zone and an inner zone. In the outer zone, convection carries energy upward. It does so in 1,000-km–size bits of gas in a manner similar to water boiling on a stove. Below the boundary, in the inner zone, radiation carries energy upward. Radiation is the process by which energy is carried by gamma rays, x-rays, ultraviolet, light (that is, visible light), infrared, or radio waves. (We usually say simply "ultraviolet" instead of "ultraviolet light" or "ultraviolet radiation" and, similarly, simply "infrared.") Gamma rays actually carry the energy in the Sun's central region. Helioseismology has told us accurately where that boundary is: about 72 percent of the way outward from the center to the surface.

Invisibly Blocked

Until recently, we could see what happened on only half the Sun at a time. That is, we could see the side that faced us, but not the other side. Of course, since the Sun rotates, some parts just rotated around the edge yesterday. But the part that will rotate around to the front side tomorrow hasn't been seen for about two weeks.

We still can't get good images of the far side, and we haven't (yet) sent a camera on a spacecraft to get such a view. But the techniques of helioseismology can be used to get fuzzy images of what active regions exist on the Sun's far side. So even a million miles of solar gas no longer stand in our way.

How and Why the Sun Shines

A hundred years ago, scientists knew that the Sun was hot, but they didn't know why. They had ideas of why, of course. They knew that when a big ball of gas contracted, it gained energy in the contraction from gravity. Perhaps this type of energy heated the Sun. They could calculate how long the Sun could live if it got its energy in this way. Its age couldn't be greater than a few million years.

But then scientists discovered by studying the Earth that rocks and fossils were billions of years old. It didn't make sense for the Earth to be older than the star around which it travels.

In the 1920s, the British scientist Arthur Eddington realized that energy could be released when atoms fused to make heavier elements. At first, Eddington and other scientists didn't know exactly how the atoms fused or which atoms did so, but they knew that nuclear fusion could release a lot of energy. It gave hope that nuclear fusion could fuel the Sun for the billions of years necessary to make the Sun older than the oldest rocks on Earth.

Fusion is the opposite of fission, in which energy is released as atoms split apart. Fission of uranium into lighter elements was identified in 1939. Fission is the process used in today's nuclear power plants.

In 1938, the physicist Hans Bethe was at Cornell University. (He was one of many scientists expelled from Germany, but he made his home at Cornell, where he was still writing scientific papers into the twenty-first

Fun Sun Facts
When the idea of nuclear fusion came up, some scientists doubted Sir Arthur Eddington's statements that it fueled the Sun. Eddington's reply? "We tell them to go and find a hotter place." Did he mean a star or Hades for the questioner?

century.) Bethe figured out certain ways in which atoms could fuse in the Sun and the other stars. Incidentally, it is so hot in the Sun that the atoms are torn apart, and it is really their central parts, their nuclei, that are fusing. So we could use the word *atom* or the word *nucleus* as we please.

One of Bethe's ideas was the *carbon cycle*. In it, an atom of carbon fuses with an atom of hydrogen. That transforms the carbon. One at a time, more atoms of hydrogen are added. Eventually, after four hydrogens have been added, we wind up with our carbon back again, but with a helium atom, too. If you compare the mass of everything that went in—four hydrogens and a carbon—with the mass of what you are left with—a helium and a carbon—you find that some mass has disappeared. It left the atoms in the form of energy.

Sun Words

An **atom** is the smallest building block of the chemical elements; a chemical element is one type of atom. Each atom contains a tiny **nucleus** in its center. The nucleus contains most of the atom's mass.

Sun Words

The **carbon cycle** is a way to fuel stars by adding hydrogens to a carbon atom and its intermediate transformations. It operates in stars hotter than the Sun. The **proton-proton chain** is the series of reactions that fuel the Sun.

In 1905, Albert Einstein showed that mass and energy are equivalent. He gave the formula for the conversion: $E = mc^2$, where E is energy, m is mass, and c is the speed of light, a large number. Einstein's formula shows us that even if only a little mass disappears, a lot of energy results. Bethe could show that the amount of energy that disappeared in the stars was enough to keep them shining for billions of years.

Though it wasn't clear at first, Bethe's carbon cycle doesn't operate in the Sun. It isn't hot enough. Instead, the *proton-proton chain* fuels the Sun. In it, protons come together, first a pair and then others adding to the result. At the end, again we have one helium atom instead of four hydrogens. And again, $E = mc^2$ shows us that we have gotten relatively a lot of energy from the disappearance or transformation of a little mass.

The Sun is so massive that it has a lot of hydrogen in it. The Sun contains 2 with 33 zeroes after it of mass as measured in grams. Since we multiply by 1,000 each time to go from one to a thousand to a million to a billion to a trillion, the Sun contains twice a billion trillion trillion grams of matter. And 90 percent of it is hydrogen. So even if only a bit of energy comes from each fusion of four hydrogens into a helium, there is still enough to fuel the Sun for 10 billion years. We are now halfway through that lifetime.

There was one legitimate objection to the idea that nuclear fusion could make the Sun shine. That is that all nuclei have positive charges. And pairs of positive charges repel each other. So how could you get two hydrogen nuclei together in order to allow them to fuse? The answer is in the incredibly high pressure at the center of the Sun, a result of all the mass above it being pulled together by gravity. And part of the answer is in the high temperature—about 15 million kelvins—at the center of the Sun. Helio-seismology has verified our ideas about the high temperature and pressure deep inside the Sun.

Fun Sun Facts

Fusion, which takes place inside the Sun, is the opposite of fission. In fission, heavy elements such as uranium and plutonium divide into less massive elements. In fusion, light elements such as hydrogen combine to make heavier elements such as helium and carbon.

If you put too much of certain kinds of uranium or plutonium together, it makes an A-bomb (atomic bomb). But if you put a lot of hydrogen together, nothing happens unless you compress it and heat it up, perhaps with an A-bomb. Only if you do so will you get an H-bomb (hydrogen bomb).

There is another big difference: Uranium and plutonium are hard to get or make. Hydrogen is very plentiful—the oceans are full of it, since water is H_2O.

Invisible Neutrinos

Though electromagnetic radiation like gamma rays or light doesn't get directly to us from the center of the Sun, one kind of particle does. This subatomic particle is the *neutrino*, a term that means "little neutral one." The neutral part means that it doesn't interact with the Sun's magnetic field. More than that, neutrinos hardly interact with matter at all.

But neutrinos are formed deep inside the Sun, where the Sun makes its energy. The temperature and pressure are so high there that nuclear fusion takes place. And in some of the intermediate reactions, as hydrogens are fused into helium, neutrinos are given off.

Neutrinos interact so seldom with matter, even with a Sun's mass of matter, that they almost all escape. Traveling at the speed of light, or almost so, they reach Earth in about 8 minutes.

Sun Words

A **neutrino** is a subatomic particle with very tiny mass and no electric charge that travels at almost the speed of light and interacts very poorly with other matter.

Fun Sun Facts

John Updike wrote a poem about neutrinos called "Cosmic Gall." He began:

> *Neutrinos, they are very small.*
> *They have no charge and have no mass.*

The latest results, though, show that the last of his points is wrong.

But if neutrinos escape from the mass of the Sun so easily, how do we detect them on Earth? The answer is: with difficulty. The basic theory was figured out by the astrophysicist John Bahcall in the 1960s. He recruited the chemist Ray Davis to work out the way to detect neutrinos. Davis thought of and perfected a scheme using chlorine. When a neutrino hits a chlorine atom, it converts to a radioactive form of argon. Davis figured out how to collect small quantities of argon and to count the number of atoms.

Davis filled a huge tank with chlorine atoms. He let it sit out for weeks or months and then counted how many argon atoms had formed. Only one argon atom had formed every other day or so! He had incredibly sensitive techniques.

Fun Sun Facts

Davis needed 100,000 gallons (400,000 liters) of chlorine in liquid form, which he got most cheaply as ordinary cleaning fluid. Dry cleaners use so much of it that it is widely available and very cheap—cheaper than milk or the bottled water so many people drink. If you worked for a company that received such an order for cleaning fluid, would you send along thousands of wire hangers?

Davis's tank is a mile underground in the Homestake Gold Mine in Lead, South Dakota. Many scientists want to have the government take over the scientific facilities, making them into a major underground national observatory. Otherwise, the owners may well close the mine and therefore prevent access to the scientific area. It is important to have some experiments, like this one, underground to keep out random nuclei coming from outer space—the so-called "cosmic rays." Cosmic rays don't pass through a mile of rock.

Davis succeeded in detecting occasional neutrino collisions through finding his argon atoms. But he detected only about a third as many as Bahcall had calculated he should. Physicists attacked, saying that Bahcall's calculations were wrong. They said that he didn't know how hot the Sun's center was. Even a small error in the temperature would have a big effect on the calculation of how many neutrinos to expect.

But Bahcall kept refining his calculation. After an original substantial correction, his predicted value has remained fairly constant for decades. And Davis's tank keeps chugging on, collecting neutrinos from the Sun. The difference between the prediction and the observations was known as the solar neutrino problem. Davis shared in the 2002 Nobel Prize in physics.

Eventually, in the 1990s, some other methods of detecting neutrinos were worked out. One sensitive method used the element gallium instead of chlorine. It was sensitive to a wider range of neutrinos, since the chlorine atoms were sensitive only to neutrinos with very high energies. Large amounts of gallium were collected in the Soviet-American Neutrino Experiment (SAGE) in what is now the Russian Federation, and in GALLEX, the gallium experiment, in Italy. The Italian gallium is held in a side-cavern in a highway tunnel deep under a mountain. The gallium experiments also show too few neutrinos to match the predictions.

> **Fun Sun Facts**
>
> The rate of one detectable interaction of a neutrino in about a day and a half is known as 1 solar neutrino unit, or 1 SNU. In a paper that brought the solar neutrino problem up to date, the great physicist Willy Fowler titled it "What SNU?"

Another type of neutrino experiment was set up in the Kamioka mine underground in Japan. It is known as Kamiokande, for Kamioka Neutrino Detection Experiment. It used a huge tank of very pure water and looked for flashes of light as neutrinos interacted with protons in the water. It not only showed the deficiency in the number of neutrinos, but it also found evidence that neutrinos could change from one type to another. The head of the project, the Japanese scientist Mashatoshi Koshiba, shared half the 2002 Nobel Prize in physics with Davis.

This evidence that neutrinos can change types, known as "flavors," is proving to be the key to the solar neutrino problem. There are three types of neutrinos. The Sun gives off only one of those types, and that is the type that the neutrino experiments have been detecting. Now it seems that en route between the Sun and Earth, the neutrinos have been changing so that they are distributed among the three types when they arrive at Earth. Since the chlorine experiment detects only one of the types, it is no wonder that it detects only one third of the prediction.

To change types, physicists tell us, neutrinos must have some mass. Massless neutrinos couldn't change. So the evidence in the successor Super-Kamiokande experiment that neutrinos change types shows that they have mass, though it doesn't tell us how much. The discovery has the potential to change important things in nuclear physics. The physicists were premature in telling the astronomers that we had our Sun temperature wrong. They were the ones who were wrong, with their conviction that neutrinos were massless. The discovery that neutrinos have mass after all could lead to major revisions of nuclear theory.

The large tubes in the Super-Kamiokande experiment sense flashes of light given off when neutrinos interact in the tank of water. Since the photo was taken, the tank had been filled, but when it was emptied and refilled in 2002, many of the photomultipliers broke and had to be replaced.

(Kamioka Observatory, ICRR [Institute for Cosmic Ray Research], The University of Tokyo)

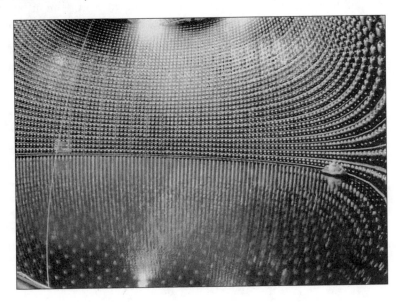

The best neutrino observatory was opened in 2001 in the Sudbury mine in Ontario, Canada. This Sudbury Neutrino Observatory uses "heavy water," water in which the ordinary hydrogens are changed to deuterium. Deuterium, known as "heavy hydrogen," has an extra, neutral particle in its nucleus in addition to the lone proton that ordinary hydrogen has. The deuterium is especially sensitive to neutrinos, and the scientists have been running various experiments to detect neutrinos as they undergo different types of reactions. They are pinning down the details of solar neutrinos.

The Least You Need to Know

- Gas has an amount of opaqueness, which we measure by optical thickness.
- Looking at the Sun through special filters or in nonvisible light can show us levels higher than the photosphere.
- Helioseismology, the study of surface waves on the Sun, tells us what the Sun is like inside.
- The Sun shines because hydrogen is fusing into helium at its core.
- The detection of solar neutrinos by several experiments has told us both about the Sun and about neutrinos themselves.

Chapter 4

The Sun Goes Up; the Sun Goes Down

In This Chapter

- ◆ Where the Sun rises and sets changes over time
- ◆ Solstices and equinoxes
- ◆ The tilt of the Earth's axis sets the season
- ◆ The Sun follows the ecliptic: the equation of time
- ◆ The Sun moves among the stars on a yearly cycle

A story is told about a grizzled veteran reporter who was next to a novice reporter at a desert location, both waiting for the night to end. The novice said, "Will the Sun rise over there in the east?" And the more experienced reporter replied, "If it doesn't, you will have yourself one hell of a story."

In this chapter, we discuss the position of the Sun in the sky as observed in different ways and how it affects our Earth.

Sunrise, Sunset

Since the time of Copernicus, Kepler, and Galileo, people have known that the Sun doesn't actually rise. Actually, of course, the Sun holds steady at the center of our solar system; it is our Earth that is turning. Only from our point of view, standing or sitting on our rotating Earth, does the Sun appear to move in the sky.

From our position on Earth, we can imagine that there is a *celestial sphere*, a large sphere surrounding us with the stars pasted on it. It appears as though that celestial sphere is rotating overhead. Actually, the Earth rotates on its axis once every 24 hours—that is, once a day. At any moment, half the Earth—the side that is facing the Sun—gets sunlight, making daytime. The other half gets no sunlight, and so it is nighttime.

> **Fun Sun Facts**
>
> A traditional story is told about what happened in 1616 after Galileo, having been shown the instruments of torture (like the rack) by authorities of the Inquisition, abjured his claim that the Earth moved around the Sun instead of vice versa. As Galileo left the room, he murmured, "And yet it moves," meaning the Earth.

The frontispiece from Galileo's book on The Dialogue of the Two World Systems, *published in 1632. This book led to Galileo's house arrest. The figures are labeled, left to right, Aristotle, Ptolemy, and Copernicus, but the figure labeled Copernicus is drawn with Galileo's face.*

(Jay M. Pasachoff)

The Sun rises in the eastern part of the sky, goes high in the sky to our south, and sets in the west. Usually, though, it doesn't actually rise exactly east or set exactly west. And it never passes exactly overhead in the continental United States. That overhead point is called the zenith, and the Sun reaches it in the United States only for people in Hawaii.

To simplify the matter, let us imagine three cases. Two of them are especially simple. First, let us put ourselves at the equator. Next, let us put ourselves at the North Pole. Finally, let us go to our actual situation, approximately one third or one half of the way between the equator and the pole. That brings us from a latitude of about 26° for the Florida Keys to 35°, which passes through southern California and Georgia, to latitudes of about 45°, which pass through Oregon, Minnesota, and northern New England, up to 50° on the central and western border with Canada.

Sun Words

The **celestial sphere** is the imaginary sphere surrounding the Earth with the stars on it. The **zenith** is the point in the sky directly over your head.

Solstices and Equinoxes

Let's go first to the equator, putting on our swimsuits. From this position, everything rises straight up. Polaris, often called the north star, lies on the northern horizon. (Polaris is near the north celestial pole, which lies above the Earth's North Pole. It is within 1° of the north celestial pole, though it isn't exactly there.) All the stars appear to move in half circles around Polaris. Near it, stars execute little circles, rising straight up and over only to set straight down. Farther south toward the east on the horizon, the half circles we see are larger. Due east, stars rise straight up, pass directly overhead (passing through our zenith), and set due west.

Just where the Sun is among the stars varies over the year. On two days during the year, known as the vernal equinox and the autumnal equinox, the Sun rises due east and sets due west. But over the year, the Sun changes the position of its sunrise, something known at least as far back as the time over 4,000 years ago when Stonehenge was built. The vernal equinox occurs around March 21. It is called an "equinox" because, geometrically, day and night are equal. (We really mean *daytime* and *nighttime*; the word *day* has different meanings: the duration of sunlight and the 24-hour period of Earth's rotation.) Actually, that calculation of an equinox is for a mathematical point at the center of the Sun. Of course, really the Sun is bigger than a point, and its top rises a little ahead of its center. That is one of the reasons daytime is a little longer than nighttime on the day of the equinox.

Star trails around the north star, Polaris. Polaris is a fairly faint star, but it happens to be located near the north celestial pole.

(National Optical Astronomy Observatories/AURA/NSF)

For three months following the vernal equinox, the Sun rises farther north on the horizon each day. It does so at a rapid rate for a while, but close to three months later, it rises less and less rapidly farther north, until it appears to pause on the horizon for a few days around June 21, which is the summer solstice. Then it starts moving south on the horizon for the next three months until about September 22, when we have the autumnal equinox.

At the equinoxes, the Sun crosses the celestial equator, the giant circle on the celestial sphere that lies above the Earth's equator. At the autumnal equinox, it is moving day by day from north to south, and it rises due east and sets due west. Only at the equinoxes does it do that. And only for observers at the equator does it go straight overhead on that day.

The units used to measure how far north or south the Sun is of the celestial equator are called declination. Declination is measured in degrees north or south of the celestial equator.

For our position at the equator, it is direct to see that the declination corresponds to the angle on the horizon north or south of the equator. Angle measured around the horizon is called azimuth. Due north has azimuth of 0°, east has azimuth of 90°, south has azimuth of 180°, and west has azimuth of 270°.

Now let us imagine ourselves at the North Pole, wearing heavy parkas. Mostly we can be standing on ice while there, though sometimes the ice opens and ships can sail up to it. While at the North Pole, the north celestial pole (marked by Polaris, the north star, nearby it) is directly overhead. The stars move in small circles high in the sky and in larger circles lower down. The circles are all horizontal. The stars neither rise nor set in the sky.

At the time of the vernal equinox, the Sun is at the equator. Since the equator lies on our horizon all around, we see the Sun all the time. As Earth turns, the Sun appears to move around the horizon but is always visible. We say that it is the "midnight Sun," since it is visible even at midnight.

For the next three months, the Sun gets higher in the sky, until it reaches an altitude (the height above the horizon in degrees) of 23½°. Then it gets lower and lower until, another three months later, at the autumnal equinox, it is again at the equator. We have had six months of sunlight. The day, if we define it as when the Sun is up, has been six months long.

Fun Sun Facts

At and near the South Pole, the Sun is set for many months each year. It is dark and, therefore, cold. But in the Antarctic springtime each year, when the sunlight hits the cold upper atmosphere for the first time in months, some of the ozone (a molecule containing three oxygen atoms) is torn apart in interactions with chlorine. That chlorine has been put into the atmosphere in Freon from air conditioners, spray cans, and other human-made sources. Use of older kinds of Freon has been banned from future air conditioners, and different propellants are now used in spray cans. Partly as a result the ozone hole may have stopped getting bigger each year. We hope that the ozone layer will recover completely, in some decades. Eventually, it should no longer ever appear. We will discuss the ozone hole further in Chapter 27.

Now let us go to our real position, at moderate latitudes. Now the whole sky seems to tilt for us. At the vernal equinox, the Sun still rises due east, but now it slants toward the south as it goes through the day. It again sets due west. On succeeding days, it rises farther and farther north, until June 21, when it rises 23½° north of due east. Then it rises farther south each day until September 22, when it is again rising due east. For the rest of the year, it rises south of due east.

Fun Sun Facts

Though the longest day of the year in terms of daylight is June 21, the latest sunset doesn't occur until June 27. The earliest sunrise takes place on June 15.

Fun Sun Facts

Why isn't the duration of sunlight equal to the amount of nighttime on the days of the equinoxes? First, as we have seen, the top of the Sun rises a couple of minutes ahead of the center of the Sun, for which the calculations are made. Second, the Earth's atmosphere bends sunlight, so when we are looking straight out, we are really seeing light that has been bent downward a bit toward us. That adds another couple of minutes. Third, the Earth's orbit around the Sun is elliptical, and the Earth's speed on its orbit around the Sun varies smoothly over the year. That adds a little more time. Daytime actually is greater than nighttime by about 10 minutes at the day of the equinoxes. It takes a few extra days around the times of the solstices for actual day and actual night to be equal. Thus, equal day and equal night occur a few days before the vernal equinox and a few days after the autumnal equinox.

Princess Summerfallwinterspring

In the early years of television, from 1947 through 1960, a favorite kids' show was based on the puppet Howdy Doody. Howdy now lives in the Smithsonian in Washington. Along with the main characters—Howdy and live people playing Buffalo Bob and Clarabelle, the clown—were some secondary characters. One of the nicest was Princess Summerfallwinterspring. Her name embodied the seasons.

But why do we have seasons? Many people think incorrectly that the seasons occur because the Earth is at different distances from the Sun at different times of year. But that idea is just wrong. Though the Earth's orbit is an ellipse, the ellipse is so close to being a circle that the difference in distance doesn't matter. Furthermore, the Sun is actually closest to the Earth on January 4 each year. That is winter for us in the northern hemisphere, not the summer that would be the result if the distance mattered.

No, the seasons are caused by the tilt of the Earth's axis. The axis stays pointing in the same direction as the Earth moves around the Sun. In the summertime, the north half of the axis is pointed toward the Sun. For us at middle northern latitudes, the Sun is more nearly overhead. It is up longer and is higher in the sky. Those factors make it hotter for us. Because the Sun is overhead, each bit of sunlight brings its full energy to the area of Earth it hits.

What is it like in the southern hemisphere when we have summertime in the northern hemisphere? The Sun is then low in the sky there and is up for a relatively short time each day. That makes it cooler for the southern hemisphere. Each bit of sunlight is spread out, since it hits the Earth at a big angle. Thus, the energy in that sunlight has to heat more area of the Earth, and the Earth is just cooler. That makes it winter for them.

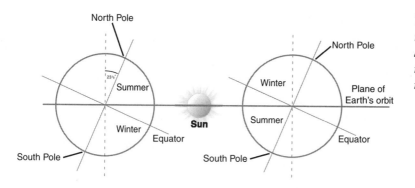

The seasons are caused by the tilt of the Earth's axis, not from changes in the Earth's distance from the Sun.

If the Earth's orbit weren't tilted, we wouldn't have seasons. Mars's orbit is (now) tilted at about the same angle as Earth's. So Mars has similar seasons. Each Martian spring, giant dust storms arise as a result of the change in seasons. The planet Uranus is lying on its side, so its seasons are very strange. Its set of seasons takes 84 years to cycle, since that is Uranus's orbital period around the Sun. The North Pole of Uranus, for example, has a midnight Sun that lasts 42 years followed by 42 years of darkness.

Figure 8s in the Sky

Ice skaters do figure 8s all the time, but did you know that the Sun does a figure 8s in the sky with respect to the stars? It does so in part because the Sun doesn't keep good time—at least, not good time compared to your watch.

Johannes Kepler showed in 1609 that the planets go around the Sun in ellipses, which was his first law of planetary motion. His second law of planetary motion was that the imaginary line joining the Sun and a planet sweeps out equal areas in equal times. Therefore, when the planet is a little closer to the Sun, it has to move a little faster on its orbit. Now, in a given length of time, as a planet moves on its orbit, the line joining the planet to the Sun sweeps out a triangle, with the sharp point of the triangle at the Sun. Whatever length of time we consider (say, a day), the area of that triangle has to remain the same. When the planet is closer to the Sun, the two legs of the triangle that link the planet to the Sun are relatively short. So the third leg, the one along the planet's orbit, has to be relatively long. A long arm, corresponds to a high speed for the planet in its orbit.

As a result, the Sun appears to move a little faster in the sky in the wintertime, since the Sun is closest to the Earth on January 4 each year. If you define "noon" as when the Sun is due south of you, then you don't have to worry about the Sun's rate. But if you carry a watch, you can discover (perhaps by putting up a stick to cast a shadow) that the Sun isn't actually due south at noon on your watch.

Because it is convenient to wear watches, the world has defined an average time, called mean solar time, for watches and clocks to keep. But on a day-to-day basis, the Sun doesn't keep that average. Sometimes it is ahead, and sometimes it is behind.

Remember also that sometimes the Sun is at northern declinations—that is, north of the celestial equator. And sometimes the Sun is at southern declinations—that is, south of the celestial equator.

The two effects together can show up in interesting ways. The following figure shows a very remarkable image made by setting up a camera in a window for a year. The photographer, Dennis di Cicco, had his camera set to take a photo at the same time of the morning, 8:30 A.M., every two weeks. You can see that the Sun is sometimes ahead of its average and sometimes behind, and how much it is ahead or behind varies at the same time the Sun's height above or below the celestial equator varies. This image shows the Sun's positions tracing out a figure eight. At certain times of the year, di Cicco, an editor of *Sky & Telescope* magazine, left the camera open from sunrise onward for a while, to show the angle at which the Sun rose.

The analemma, shown as a series of actual photographs taken at the same time of day every 10 days for a year. The Sun was photographed through a dense, special solar filter; and on one day, the background was photographed without a filter. On certain days, the camera was also left open to expose the Sun rising.

(Dennis di Cicco)

In the preceding figure, the summer solstice shows the Sun at the highest point, at upper left, and the winter solstice shows the Sun at the lowest point, at lower right. This figure-8 shape is known as the *analemma*, and you sometimes find an analemma drawn onto a globe, for convenience. It is often put in the middle of the Pacific Ocean, just because there is a lot of blank space there.

The Solar Scoop _____

You can see analemmas on the Internet at the Astronomy Picture of the Day, antwrp.gsfc.nasa.gov/apod for July 9, 2002, and for March 20, 2003, using the following: antwrp.gsfc.nasa.gov/apod/ap020709.html, for 2002 July 09 and antwrp.gsfc.nasa.gov/apod/ap030320.html for 2003 March 20.

At the widest point of the analemma, the Sun is about 16 minutes ahead or behind mean solar time (see the preceding figure). Some of the difference comes from the varying speed of the Earth in its orbit around the Sun, and the rest comes from the tilt of the Earth's axis.

If you look at a sundial, you will find that the time on the sundial differs by as much as 16 minutes from the time on your watch. But if you make the shadow stick (known as the gnomon) on a sundial in the shape of an analemma, you have a sundial that keeps time and that matches time on your watch, and you can read it to a minute or so of accuracy.

The comparison of the apparent time kept by the actual Sun with the mean solar time kept on your watch is called the equation of time.

Fun Sun Facts

Most people are familiar with four basic time zones for the United States: Eastern, Central, Mountain, and Pacific. But the Virgin Islands are in the Atlantic Time Zone and Alaska, Hawaii, and Samoa have their own time zones. A group of South Pacific islands, including Guam and the Northern Marianas, are in a time zone fairly recently (in 2001) named Chamorro. So there are eight United States time zones.

Russia is so big that it spans nine time zones. How many time zones does China have? The answer is one. Everyone in China is on Beijing time, even though China is so broad.

Not all time zones are spaced at hourly intervals. The central part of Australia is only a half hour off from time in eastern Australia. That makes a gap of 1½ hours to the time zone of western Australia. Nepal's time zone is only 15 minutes off from India's. King Edward VII of England kept time a half hour different from Greenwich mean time at one of his estates.

In the Zone

If you kept time by the actual Sun, then somebody a few miles to the west of you would have noon slightly later than you do. These small effects didn't matter until the growth of the network of trains in the nineteenth century meant that it would be good to have a standard set of time zones. The time zones were set up at an international conference in 1884. Within each time zone, the time would be the same. Since the Sun goes 360° around Earth in 24 hours, it goes 15° in 1 hour. Therefore, each time zone was made one hour wide. The time zones are spaced with their centers at 15° intervals.

Saving Daylight

If you live in Hawaii, the Sun sets at about 6 P.M. all year, a little later, but not very much later, in the summer. But if you live at temperate latitudes, the length of the day varies a lot more.

Let's say that as you go into April, you notice that the days are getting longer. It is lighter longer after you get home from work or school. To give yourself more sunlight hours, then, why not just take 7 P.M., when it gets dark, and call it 8 P.M.? So, if you get home at 6 P.M. from work or school, you now have two hours of sunlight (6 P.M. to 8 P.M.) instead of only one. Businesses can then sell a lot of extra items, like barbecue briquettes. But if you are a farmer who works until sundown, then your after-work interval until bedtime is likely to be shorter.

This kind of transition is called daylight saving time and, in European countries, "summer time." Most places in the continental United States follow daylight saving time, adding an hour in the spring and taking an hour away in the fall.

By moving the clock forward in the springtime, it means that 6 A.M. is now 7 A.M. In the fall, when you move the clock back, 7 A.M. becomes 6 A.M., and kids waiting for schoolbuses might have to stand in the dark. So some people are opposed to daylight saving time.

Hawaii doesn't follow daylight saving time because the length of the day doesn't vary much at its latitude. Alaska doesn't follow daylight saving time because so much of it has midnight Sun. Much of Indiana, the part on Eastern time, doesn't follow daylight saving time. The opening show of the TV series *The West Wing* in 2002–2003 revolved around some of the characters getting stranded because they didn't realize that they were switching time zones as they traveled in Indiana.

Some places are on double daylight saving time. Two extra hours of daylight can save a lot of oil.

Keeping the Calendar

Once around the Sun for the Earth makes a year. But while we sit on the Earth as it travels, Earth spins on its axis. And each time it spins, it is one day more. For the moment, let's take the grand point of view, looking down from high above the solar system, out among the stars. From that point of view, when the Earth spins 365 times, it isn't quite back to its one-year position in its orbit. Earth has to spin about another quarter of the way before it gets back to the position that marks the passage of one year.

Thus, our year is 365 days long. But thousands of years ago, people didn't know so much, and tended to keep track of time more with the moon. (That is why we have the name month, which comes from "moonth." The Moon goes around the Earth around every 28 days, pretty close to the 30 or 31 days of our usual month.)

About 4,000 years ago, the Babylonian year had 360 days, and they occasionally added an extra month, to make the number of days average 365. Today's Jewish calendar similarly adds 7 leap months every 19 years.

Back in Roman times, the number of days in a year wasn't standardized. The emperor Julius Caesar called for a 445-day year in 46 B.C.E. to bring the dates of the year in line with the seasons. The Julian calendar, which was also part of his reform, introduced our basic system of leap years. The Julian calendar's year was 365 days long most of the time, but every fourth year, an extra day was added. That made the average year 365¼ days long.

Actually, however, the year is about 11 minutes short of 365¼ days. So after centuries, the calendar drifted out of synch with the seasons again. The sixteenth-century Pope Gregory XIII convened a group of astronomers and others to carry out calendar reform. He particularly wanted Easter to be realigned with the spring. The Pope, acting on the recommendations, declared that the day after October 4, 1582, would be October 15, 1582. The Gregorian calendar continued the system of leap years. But it omits the leap year every hundredth year, to make the year a little shorter. And that corrects by too much, so it restores the leap year every four hundredth year. That is why the year 2000 was a leap year, even though 1900 was not. This reform will keep the calendar in step with the seasons for thousands of years.

Fun Sun Facts
It was very confusing for people during the 10 days that were skipped on the calendar. "Give us back our fortnight," was the cry. Of course, people didn't really grow older by 10 days; it merely appeared that way on the calendar.

Though many countries adopted the Gregorian calendar in 1582, Britain and its American colonies didn't adopt it until 1752. At that time, we skipped 10 days—the day after September 2 was September 13. The calendar, which we adopted then, also moved the first day of the year back from March 25, about the equinox, to January 1. Until then, dates in January, February, and most of March would have been in one year in the Julian calendar and another year in the Gregorian. If you look at George Washington's family bible, which is preserved at Mt. Vernon by the Ladies' Association of the Union, you see his birth recorded as "11th day of February 1731/2." Skipping the 10 days made his birthday February 22 for us, and we use 1732 as the year.

The Least You Need to Know

- The Sun rises as much as 23½° north or south of east, depending on the time of year.

- On about March 21 and September 22, the Sun goes from one side of the equator to the other, making the equinoxes.

- Day and night aren't quite equal on the days of the equinoxes.

- The seasons are caused by the tilt of the Earth's axis, not by the Earth's changing distance from the Sun.

- Time zones were set up to standardize time within bands hundreds of miles wide, since the Sun is usually ahead or behind time on your watch.

Our Sun: Looking Good

In This Chapter

- ◆ What color is the Sun?
- ◆ Why is the sky blue, and why are sunsets red?
- ◆ What is the green flash and how does the Sun make rainbows?
- ◆ When does the Sun cause static on your radio?
- ◆ What is the Sun's spectrum like?

The Sun is too bright to safely look at it directly. We know, nonetheless, that it has its own color. Furthermore, it causes lots of colors around the sky. In this chapter, we discuss several aspects of the shining Sun.

What Is White?

The human eye is a wonderful sensor. When it sees a white page, it teams up with your brain to make it seem white, even when its color changes slightly. For example, light at sunset is reddish, but a notebook page that you hold up in the air will still look white to you.

The Sun gives off light all across the spectrum. The shortest wavelengths of light that we perceive are violet, somewhat longer wavelengths are blue, middle wavelengths are yellow and green, and long wavelengths of light are orange and red. It is nice to think of the friendly fellow known by the acronym of ROY G BIV: red, orange, yellow, green, blue, indigo, and violet.

If you measure with an electronic meter, you would find that the Sun gives off most of its light in the yellow and green. It gives off less light to shorter and to longer wavelengths. This mixture of all the colors just as they are coming from the Sun is known as white light. It isn't exactly the same as the mixture of red, green, and blue light coming from three projectors that, when they overlap, also look white to the eye.

The human eye detects the finest detail and the faintest images using structures called rods located in the back of the retina. It detects color using the cones. These cones come in three types—ones sensitive to red, ones sensitive to green, and ones sensitive to blue. So if the cones are stimulated in the right proportions, your eye and brain perceive white.

Fun Sun Facts

In astronomy, when looking at a faint object in the sky either directly or through a telescope, you don't usually get strong enough light to activate your eye's cones. Only the rods operate, so you don't usually see color in the sky. You certainly don't see the bright colors that show in long exposures with telescopes.

To look at the Sun directly or with a telescope, you need a filter that reflects or absorbs all but about 1 part in 100,000 or 1 part in 1,000,000 of the incoming sunlight. The best of these filters have thin metal coats deposited on glass. Many satisfactory filters have metal deposited on a plastic substance known as Mylar rather than glass. Whenever you use such a filter, you should be careful that there aren't any pinholes, tears, or other defects that allow full sunlight to leak through.

These solar filters have approximately neutral density—that is, they are roughly equally dense to all colors. A safe neutral density (that is, ND) solar filter must be at least ND 5—that is, it must absorb all but 1 part in the number 1 with five zeros (100,000). And this neutral density must extend not only across the whole visible spectrum, but also into the infrared. Photographic filters, such as the Wratten series made by Kodak for use with cameras, aren't very dense in the infrared and so should not be used for looking at the Sun with your eye, even if they say ND 5.

Many of these filters aren't strictly neutral. They may let slightly more blue light through, giving the solar image a blue tinge. Many people find this effect less pleasing than that of the metal-deposited filters, which often have a slight orange tone.

Why Is the Sky Blue?

When light bounces off tiny particles, we say that it is "scattered." Its direction changes, and it can bounce around. But the shorter the wavelength, the more effectively light is scattered. So blue light scatters more effectively than red light.

This kind of scattering off small particles compared with the wavelength of light is known as Rayleigh scattering. Lord Rayleigh (1842–1919), in England, worked out the theory in the nineteenth century. (Later, he got the 1904 Nobel prize for his work on gases, including the discovery of argon.)

The dependence of Rayleigh scattering on wavelength is very strong. It varies inversely with the fourth power of the wavelength. Red light at 650 nanometers (650 billionths of a meter) is $650 \div 400 = 1.6$ times longer than blue light at 400 nanometers. But it scatters better by a factor of 1.6^4. You can do this math in your head by noting that $1.6^4 = 1.6^2 \times 1.6^2$. Since 1.6^2 is about 2.5, and 2.5^2 is about 6, we see that blue light scatters better by a factor of about 6 even though it is only a factor of 1.6 times shorter.

When the scattering particles get bigger compared to the wavelength of light, they don't cause Rayleigh scattering any more. Then they scatter evenly across the spectrum. So though the particles that make the sky blue are small, larger particles merely spread out the sunlight in various directions. The larger molecules of water vapor that make up clouds spread out the sunlight evenly. That is why clouds are white.

Why Are Sunsets Red?

When you look at a sunset, you are looking low on the horizon. The sunlight has traveled through much more air to get to you than it does when the Sun is overhead. As the sunlight travels through the air, it undergoes Rayleigh scattering. At sunset, that makes the sky blue for the people on the Earth between you and the setting Sun. Their blue skies subtract so much sunlight that what is left appears reddish.

If you have ever taken a photograph at sunset, you know that it turns out with a reddish cast. Sometimes that cast can be taken out in processing, but sometimes it is too strong. Still, the human eye and brain team up to see the sunlight as white even when it is really reddish.

Sunsets are red because the blue light has been scattered to make blue skies for other people.

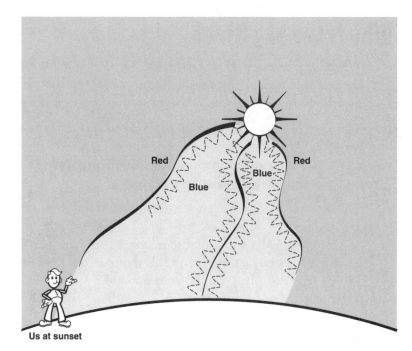

Us at sunset

What Is the Green Flash?

The green flash is real, though many have doubted it. Why, people say, should the Sun ever look green on the horizon? The answer lies in the bending of sunlight and in its absorption.

The Earth's atmosphere is a thin layer that curves around Earth. As sunlight passes through it at a low angle, the sunlight is bent. But the amount of bending depends on the wavelength. Violet light and blue light are bent the most, and red light is bent the least.

If you then look back at the Sun, you might, in principle, see a series of overlapping round images. The blue image would be bent the most, so when you look back along it, it would appear the highest. The red image would be bent the least, so when you look back along it, it would be relatively close to the real geometric direction to where the Sun would be if the Earth didn't have an atmosphere.

But of that vertical series of Sun images, little of the violet and blue light reaches you because it is scattered out to make blue skies elsewhere. You thus have only ROY G. When you look low in the sky, the amount of water vapor accumulates. Water vapor absorbs orange and yellow light very well. That leaves you two overlapping images: a higher one of green and a lower one of red. As soon as the red one sets, you are left with only the green. It lasts just a few seconds, and at that point just a green rim appears at the top of the Sun. Sometimes distortion in the atmosphere makes that green rim appear to separate a bit from the top of the solar image. In any case, that green bit flashes into visibility for a few seconds.

The effect is best seen when looking out over water because then your horizon is perfect. Of course, there can't be any distant clouds that block your view of the setting Sun. Don't stare too long at the Sun at any time, including while you try to see the green flash. You don't want to injure your eyes, and, on a lesser ground, you don't want any green that you may see to be just a negative reaction in your brain to the brighter red image that has set.

A green flash can occur at sunrise as well. However, usually we don't stand out at sunrise waiting for the first glimpse of the rising Sun, and we usually don't know exactly where on the horizon to look.

Rainbows

Rainbows can be fantastically beautiful. In certain places—like Hawaii—they are very common. In other places, they are rare wonders. They result from sunlight bending and being reflected inside water droplets.

Imagine what happens when a light beam enters a raindrop. Some of the rays hit the front surface of the raindrop at an angle, so they bend as they enter. Any such bending depends on wavelength. Again, the blue rays bend more than the red rays. So, already within the raindrop, the different colors are heading in slightly different directions.

Once inside the raindrop, the light rays reflect off the inside back of the drop. Each ray, no matter what its color, reflects off at the same angle at which it hits. But since the different colors were coming in slightly different directions, the colors are still spread out after the reflection. And the bending of the different rays by different amounts as they leave the raindrop's front surface doesn't bring the rays together. They remain dispersed into the rainbow that we know and love. The band of red always lies above the band of yellow, which lies above the band of blue, and so on.

Fun Sun Facts

In 1637, René Descartes figured out how the shape of a rainbow is formed by internal reflection in raindrops. Later, Isaac Newton figured out how the colors of the rainbow are formed, incorporating his understanding of how refraction forms a spectrum. He published his ideas in 1704 in his book titled *Optics*.

If you work out the details, it turns out that the raindrops reflect the dispersed sunlight back to you in a circle about 42° across around the point opposite the Sun. Of course, we see rainbows only when the Sun is above the horizon, so the point opposite the Sun is below the horizon. That is why we don't see a whole rainbow; we see only the part that peeks up over the horizon. When the Sun is low in the sky, the rainbow extends higher up in the opposite part of the sky.

The rainbow is formed by sunlight bending as it enters and leaves a raindrop while, in between, bouncing off the back of the raindrop. The bounce is the same for light of all colors, but the amount of the bending is different for the different colors.

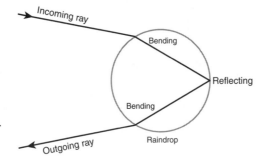

Fun Sun Facts

Seeing a rainbow is nice, and seeing a double rainbow can be even more rewarding. A double rainbow contains a second bow outside the first bow. To make the first bow, the light rays bounce once within each raindrop. To make the second bow, they bounce twice inside each raindrop. As a result, the order of the colors is reversed in the secondary rainbow. It is about 51° from the point opposite the Sun, but it is fainter than the primary bow.

Dogs and Pillars in the Sky

Sometimes the Sun causes rainbowlike effects in the sky that aren't from water droplets. They often come from ice crystals high in the atmosphere. Sun dogs and Sun pillars are formed in that way. Sun dogs are to the sides of the Sun, while Sun pillars are above or below the Sun.

Sun pillars form because the ice crystals that are bending and reflecting sunlight are flat. They are eight-sided, like snowflakes. As the ice crystals fall to Earth, the air resistance keeps them floating with the flat side parallel to the ground. For this reason, they are aligned enough to cause the effect of a pillar of relatively bright light, most often seen extending upward from the setting Sun. Sun pillars are the same color as the Sun, and they, therefore, can be reddish near sunrise or sunset. They can also be seen between the Sun and the ground. See, for example, antwrp.gsfc.nasa.gov/apod/ap010313.html.

Other shapes of ice crystals lead to other atmospheric effects involving the Sun. Sometimes you see a complete or partial halo around the Sun. This halo is most often 22° from the Sun, which is just greater than the width of your open hand held at the end of your outstretched arm. The inner edges of these halos are reddish.

When the ice crystals include flat ones, such as the ones that make pillars, you may see bright regions of the halo to the left and to the right of the Sun at the same altitude as the Sun. These bright spots, which can show rainbows of color, are known as sun dogs.

Static May Fade "Here Comes the Sun"

We see the Sun with our eyes using the light it gives off, but the Sun gives off radio waves, too. Whereas the sunlight is fairly steady, the radio emission from the Sun varies wildly. Giant solar flares, for example, give off bursts of radio waves that can overwhelm reception of radio stations on Earth. These flares are more common during the maximum of the sunspot cycle and in its declining phase, and powerful flares almost never occur during sunspot minimum.

Various types of radio bursts occur. Some chirp in frequency while others are more uniform across the spectrum of radio frequencies. Many of the bursts can be traced to sources high in the solar corona, since certain frequencies don't penetrate from lower down.

Radio solar observatories are rarer than optical ones, but they do exist. At Nobeyama, Japan, there is a network of dozens of small solar radio telescopes. Using computer calculations, the observations taken simultaneously with all these instruments are used to make a picture—visible to the eye—of the radio Sun. But if you just listen to the input of any one of those telescopes, what you hear sounds like just static.

Rainbows with Lines

If you spread out sunlight with a tiny prism as the light comes into your window, you see a rainbow of color. About 200 years ago, in 1802, the physicist William Wollaston thought he saw gaps in the colors as they changed from one color to another. But he reported only a handful of such gaps.

By 1814, the German optician Joseph Fraunhofer had perfected a device, called a spectroscope, for spreading out the solar spectrum. One of his improvements on previous work was to take the spectrum not of the whole Sun but only of a narrow band across it. Only the sunlight that went through a (say) vertical "slit" was spread out into a spectrum. In such a case, we could see the spectrum of that one line of sunlight. You can imagine that the spectrum of an adjacent line of sunlight would be a similar rainbow, but displaced to the side by a slit width. By the time you took the spectrum without any slit, you would have so many overlapping rainbows of color that any detail in the color would be lost.

With Fraunhofer's spectroscope, he could study the spectrum of the Sun in detail. When he spread out the spectrum from side to side, he found that there were a handful of vertical dark lines across the spectrum. These dark lines correspond to colors at which less light is received from the Sun than at adjacent colors. We now call them absorption lines or Fraunhofer lines.

In general, specific colors on the spectra of astronomical or other objects that are brighter or fainter than neighboring colors are called spectral lines. When the specific colors are relatively faint, we have *absorption lines*, which are known as Fraunhofer lines when they occur in the visible part of the Sun's spectrum. When the specific colors are relatively bright, we have what are called *emission lines*. Basically, stars, including the Sun, have only absorption lines.

Sun Words

Absorption lines are locations where specific colors of a spectrum are relatively fainter than adjacent regions on either side. In the visible part of the solar spectrum, they are known as Fraunhofer lines.

Emission lines are locations where specific colors of a spectrum are relatively bright.

Fraunhofer labeled his lines starting with the red end using the letters A, a, B, C, D, E, b, F, G, and H, and he used I for the blue end of the spectrum. We still use a few of these notations. The C-line was subsequently found to come from hydrogen and is better known as H-alpha, since it is the first line (Greek alpha) in a series of hydrogen lines in the visible spectrum. Fraunhofer had noted that the wavelength of D was the same as that of sodium in his lab. On higher resolution, the D-lines turned out to be a pair of lines close together; they are known as the sodium D-lines. The H-line turns out to be an especially strong line and to arise in calcium that has been heated enough to lose one of its electrons. On Fraunhofer's original drawing, of which a hand-colored version exists in the Deutsche Museum in Germany, it is already clear that there is an almost equal dark line just to the blue of the H-line. Sometime in the nineteenth century, it began being called the K-line. We now know that both the H- and K-lines are caused by ionized calcium. We also still talk of the magnesium b-lines.

These spectral lines turn out to be the key to understanding the surface of the Sun. From which ones are there—and many or most turn out to be from iron—we can tell what chemical elements are on the Sun and in what relative abundances. We can also take the Sun's temperature with them. These Fraunhofer lines also exist on other stars, and study of the lines has led to the understanding of the distant stars.

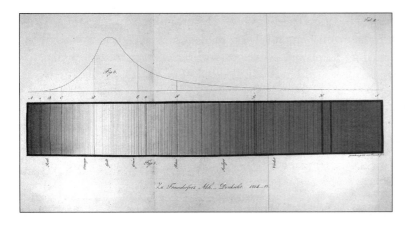

The original solar spectrum discovered by Joseph Fraunhofer in 1814, with his labels for the darkest lines that crossed the spectrum from top to bottom. The top of the diagram shows the overall brightness of the Sun at different colors.

(Jay M. Pasachoff)

Planets from Afar

In 1600, Giordano Bruno was burned at the stake in Florence, in part for talking about the "plurality of worlds." But finally, by 2000, astronomers had discovered worlds beyond our solar system, and we now know of more than 100 of them.

The discovery of most of these "exoplanets," or extrasolar planets, depends on the Fraunhofer lines. These lines are known to be at precise wavelengths. When they are observed at slightly different wavelengths, astronomers know that the gas that formed them is moving toward or away from them. When gas moves toward you, the spectral lines that it causes are shifted toward the blue end of the spectrum. When gas moves away from you, the spectral lines that it causes are shifted toward the red end of the spectrum.

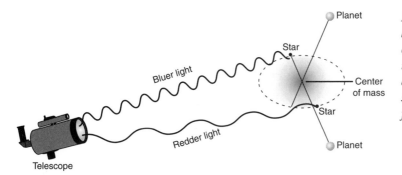

More than 100 planets have been discovered around stars other than the Sun by detecting and studying the periodic changes in wavelengths of spectral lines in the light from the stars.

If you heat the gas iodine, it absorbs incoming sunlight in particularly sharp spectral lines—that is, the lines are at colors (wavelengths) that are determined very precisely. Such sets of spectral lines were compared with the spectral lines coming from distant stars. The measurements were made over periods of years, especially by a group of astronomers in Switzerland and a competing group of astronomers in the United States. One day in 1995, the electrifying announcement came from the Swiss astronomers: A planet had been found around a distant star. The star, from its Fraunhofer lines, had been measured to go to and fro with a period of four days. From basic physics, including the third law of Isaac Newton, we thus knew that there was something orbiting that star that was going fro and to. Gravity was causing the effect. From the amount of the to-ing and fro-ing, the astronomers could get lower limits on how massive the planet must be. Similar methods were then used by both groups, and then by others, to discover the many dozens of planets of which we now know.

The Least You Need to Know

- The Sun gives off white light.
- The sky is blue because of light scattering that depends strongly on wavelength, leaving sunsets to appear red.
- The green flash occurs because of refraction and absorption of sunlight.
- The Sun gives off radio waves that cause static.
- The Sun's spectrum is a continuous rainbow with missing colors.
- We detect exoplanets by the manner in which spectral lines of their parent stars vary in wavelength.

Part 2

The Sun Through Time

The Sun is our nearest star. How does it resemble the others? What were our baby pictures like, and how do we learn about how our star grew up? Never mind that our star is a dwarf. Five billion years of growing up should give us maturity. Anyway, we are not unique. In some sense, our Sun is a clone. We started figuring out the role of the Sun in the universe when the greatest figures in the history of astronomy—including Copernicus, Galileo, and Kepler—put their minds to figuring out how the universe is laid out. Within the last decade, over a hundred different planets have been found around over a hundred different suns. It's fun to know how we know. Maybe it's not so much fun to find out how our Sun will die—but we've got lots of time before that happens.

The Sun as a Star

In This Chapter

- ◆ The Sun formed from gas and dust
- ◆ The Sun is mainly made of hydrogen
- ◆ The Sun is a star
- ◆ We learn what the Sun is made of from its spectrum

Annie Oakley sang, in the Irving Berlin musical *Annie Get Your Gun*, "I've got the Sun in the morning and the Moon at night" and "moonlight gives me the Milky Way." In fact, the stars in the Milky Way are distant suns, and our Sun is just a close-up of what many of the distant stars look like. Let's discuss how the Sun got to be as it is and how it compares with other stars.

Presto, the Sun

On a dark night, far from city lights, you can see a faint band of brightness arch across the sky. This Milky Way becomes more apparent to the eye the longer you are outside, shielded from lights that ruin your night vision. The Milky Way we see is composed of billions of stars as well as gas and dust between the stars. The processes that formed our Sun are still going on, and new stars are forming all the time.

About five billion years ago, a cloud of gas and dust began to fall together. Gravity is always pulling everything toward everything else, and perhaps some random fluctuation in the cloud of gas and dust made it denser than its surroundings. Once gravity starts pulling gas and dust together in this way, very little can stop the free fall.

How did it all begin? Maybe a cloud of gas and dust started out randomly to form the Sun, but maybe a supernova nearby in space started the collapse. Astronomers find some radioactive aluminum that may be left over from that supernova. The question is controversial.

When you take a random cloud of gas and dust, it may have a little spin in some direction. It would be odd if it were completely not spinning. When something that is spinning collapses, it spins faster. The principle involves a concept called angular momentum. The amount of angular momentum that a system of matter has depends on how fast it is spinning and how far the various spinning parts are from the central axis. The faster something is spinning or the farther spinning parts are from the axis, the higher the amount of angular momentum is.

The amount of angular momentum that any system has remains constant, unless some outside force changes it. A merry-go-round doesn't just stop abruptly. Brakes and gears have to slow it down, and they apply outside forces that stop the merry-go-round's spin.

As the solar system collapsed, it had to spin faster because its various parts were closer to its axis. The process defines directions in the spinning cloud: along the axis or perpendicular to the axis. Along the axis, there is no spin, and the cloud simply continues to collapse, getting smaller and smaller. But perpendicular to the axis, the matter goes around and around the axis in circles or ellipses. Let us consider matter in a flat plane perpendicular to the cloud's axis of spin. Its forward momentum tends to keep the matter going straight ahead, even while gravity is pulling it inward. Being thrown forward by the forward momentum keeps the matter from collapsing as fast. We often say that "centrifugal force" is throwing things outward, but physicists accurately say that there is really no such thing as centrifugal force. It is a fake force, just acting as an explanation for why things don't move inward as fast as you might expect without considering everything's inertia from forward motion.

We can think of the cloud of gas and dust as a set of spinning disks. The disks from the top and the bottom are being pulled by gravity toward the central disk. But the disks themselves don't collapse as fast. Within each disk, since its angular momentum doesn't change, it spins faster as it gets smaller.

A Hubble Space Telescope view of part of a cloud of gas and dust known as the Eagle Nebula. Stars are forming inside its pillars, one of which is shown here. The pillars are known as EGGs, or Evaporating Gaseous Globules. Radiation from the new stars forming at their tips evaporates the material that has shielded them long enough to allow them to form inside the cocoons of dust.

(Jeff Hester and Paul Scowen, ASU and NASA)

Fun Sun Facts

The concept that angular momentum doesn't change affects not only the solar system, but also ice skaters. If you watched Sarah Hughes's gold-medal performance at the 2002 Winter Olympics, you could see that she started each spin with her arms outstretched. Then she pulled her arms inward to make herself spin faster. The laws of physics were making her spin faster to keep the amount of her angular momentum constant. Because her body parts were closer to her axis of spin, she had to be spinning faster.

We wound up with a spinning disk of gas and dust. The central region had the most material and became hotter as energy was released from the force of gravity. This hot gas in the center became hot enough for nuclear fusion to begin. At that instant, the Sun was born.

In the disk around this young Sun, clumps of gas and dust themselves began to coalesce. We say that they became planetesimals. Eventually, the planetesimals began to clump into larger objects. The largest of these objects are the planets.

The young, hot Sun threw off particles into space. They expanded outward in a fierce "solar wind." This solar wind tore away most of the gas and dust in the inner part of the solar system. The energy from the young Sun heated the rest and evaporated the dust. We were left, about five billion years ago, with the solar system we know: a shining Sun at the center with planets orbiting it.

Whither the Sun

When we manage to measure how many atoms of each type there are, the answer is somewhat surprising. Though we on Earth see lots of silicon in the rocks and iron in other objects and breathe air that contains mainly nitrogen and oxygen, these elements make up together less than 1 percent of the atoms in space. It turns out that about 90 percent of the atoms in space are hydrogen, another 9 percent or so of the atoms are helium, and all the other 100-plus types of atoms make up less than 1 percent.

What are these elements? They are made of subatomic particles called protons, neutrons, and electrons. Each atom is made of a nucleus with a positive charge surrounded by electrons that have negative charges. We ordinarily think of atoms on Earth as being in a neutral state. They have as many negative charges as positive charges, so they are electrically neutral. The nucleus has as many charges in it as it has protons. The neutrons in the nucleus, as their name implies, are neutral.

In many places in the universe, the atoms are not whole. An atom that has lost one or more electrons is called an *ion* (pronounced *eye'on*). An atom that has lost one electron is "once ionized." An atom that has lost two electrons is "twice ionized," and so on. The hotter a gas is, the more of its atoms are ionized.

In the center of the Sun, the atoms are completely ionized. The nuclei and the electrons are floating freely. A gas that is ionized like this is called a *plasma*, not to be confused with the plasma in human blood. Most of the universe is made up of plasma. Often people speak of three states of matter: solid, liquid, and gas. Informally, plasma is sometimes called a fourth state, though it is really a type of gas. We often distinguish plasma because its components have electric charges. Therefore, they can be bent by magnetic fields.

Sun Words

An **ion** is an atom that has lost one or more of its electrons. A **plasma** is a gas composed of ions and the electrons that balance the charge.

The simplest element is hydrogen. Ordinary hydrogen has just one proton. In a neutral hydrogen atom, that proton is surrounded by one electron. In the center of the Sun, when we talk of hydrogen, we really mean that proton alone.

Protons are about 1,800 times more massive than electrons, so the nuclei dominate the masses of atoms. Neutrons have about the same masses as protons, though they have no electric charge. Only certain combinations of protons and neutrons are stable.

A certain form of hydrogen contains not merely one proton, but also one neutron. This type of atom is called heavy hydrogen or deuterium. The nucleus itself is called a deuteron. It is very stable, and about 1 part in 6,000 of all the hydrogen atoms in the oceans on Earth are deuterium.

Fun Sun Facts

Since deuterium undergoes fusion at lower temperatures than ordinary hydrogen, many scientists look toward deuterium as the fuel for fusion reactors. The fuel would then be obtainable cheaply from ocean water. Research on building such reactors, though, is such that power from fusion is about 50 years away. Cynics say that power from fusion has long been and always will be about 50 years away.

A third form of hydrogen, known as tritium, combines two neutrons with one proton. Tritium is not stable; after some time, it spontaneously disintegrates. We say that it is radioactive, a word coined by Marie Curie about 100 years ago. Her thesis on radioactivity led to the first of her two Nobel Prizes, which she shared with her husband, Pierre (and with Henri Becquerel).

When an atom has two protons instead of one, it is helium instead of hydrogen. Stable forms of helium exist with one or two neutrons. Helium-3 has two protons and one neutron, while helium-4 has two protons and two neutrons. Helium-4 is the ordinary and dominant kind.

Such variations among the elements, with a fixed number of protons but different numbers of neutrons, are called *isotopes*.

In the first 3 minutes or so after the Big Bang, some 15 or so billion years ago, hydrogen (including some deuterium) and helium were formed, along with very small amounts of the lightest elements like lithium. Carbon, oxygen, and other heavy elements up to iron are formed in stars. The heaviest elements are formed in stellar explosions, as we shall see later in Chapter 10.

Sun Words

Isotopes are forms of atoms with different numbers of neutrons; the neutrons add mass but not charge. For example, the most common isotope of hydrogen has just a proton, and another isotope of hydrogen is deuterium, which has a proton plus a neutron.

When the Sun formed, it incorporated the gas in space, which was 90 percent hydrogen, 9 percent helium, and less than 1 percent everything else. Deep inside, it is now cooking some of its hydrogen into helium. As we saw in the previous chapter, energy results and makes the Sun shine. At the rate that this process happens, the Sun is about halfway through its 10-billion–year lifetime.

> ### Fun Sun Facts
>
> In 1929, the young Harvard astronomer Cecilia Payne's study of spectra showed that most of the gas in the atmospheres of stars is hydrogen. This was a surprising result and was so novel that Payne was pressured not to publish it. It took a few more years—and work by such people as Henry Norris Russell and Donald Menzel—to verify Payne's results. Payne, using her married name of Cecilia Payne-Gaposchkin, went on to later triumphs. For the last few years, there has been much discussion of whether her gender impeded her recognition for her discovery. In any case, she is now widely acknowledged as the pioneer.

The actual process that fuels the Sun is known as the proton-proton chain. In it, two protons interact to form a deuteron. Then the deuteron interacts with another proton to form helium-3. At the same time, an adjacent helium-3 is being formed out of a deuteron (again formed out of two protons) and a proton. Six protons in all have gone into this mix. Then the two helium-3s combine, and the result is helium-4 plus two protons. Though the mass is about six times the mass of a proton, it is slightly less than the total mass of the six protons that went into the element formation. The difference in mass is released from the atom in a quantity measured by $E = mc^2$, the formula discovered by Albert Einstein as part of his special theory of relativity in 1906. In the previous chapter, we saw how the discovery of neutrinos from the Sun, which led to the Nobel Prize in Physics in 2002, has verified this proton-proton chain.

The other elements in the Sun are measured in a couple of ways. Examination of the Sun's spectrum is a major method. The telltale fingerprints of almost all the different elements can be found in the solar spectrum and can be analyzed to tell the quantities (abundances) of the elements. Another way is to study meteorites, rocks of stone or iron/nickel that land on Earth. They can be analyzed in laboratories, and the abundances of the elements can be measured relative to each other. Though the hydrogen and helium have escaped from the meteorites, the results for the other elements agree.

The Sun Is a Star

When you examine the rainbow of color from the stars, using a telescope and a device called a spectrograph, the resulting spectra have telltale differences. Overall, however, the spectra we get from the stars are similar to the spectrum we get from the Sun, though they are much less bright. Even the brightest star is 10 billion times fainter than the Sun.

The spectrum of each star, the Sun included, is a rainbow of color crossed by some gaps in the color. (See page 2 of the color insert.) For many types of stars, the more carefully you look at the gaps, the more there are. Under high resolution—that is, high magnification both in the instrument and in your viewing or recording the result—the gaps look like narrow, dark lines crossing the spectrum. They are thus known as spectral lines.

In the previous chapter, we saw how, in 1814, Joseph Fraunhofer mapped out the spectral lines from the Sun in detail for the first time. These Fraunhofer lines exist in all types of stars, though they often take on different intensities or patterns than the Fraunhofer lines on the Sun.

Spectra Reveal Stars' Secrets

One key to understanding the spectral lines is the set of lines from hydrogen. This set makes a distinctive pattern, with a line in the red, a second line in the blue, and more lines getting closer and closer as you look in the violet. Such a pattern is known by mathematicians as a converging series. Whenever you see such a set of spectral lines with this pattern, you know that you are seeing hydrogen gas. This set is known as the Balmer series.

Fun Sun Facts

Some decades ago, scientists using the largest telescopes discovered faint lines in some enigmatic pointlike objects in the sky. When a few of these lines became visible, their converging pattern conveyed that they were from hydrogen. They were shifted in wavelength because the source was so far away and, therefore, was moving rapidly in our expanding universe. If they are so far away, they must be exceptionally bright for us to see them. Only giant black holes with billions of times the mass of the Sun explain them. They are known as quasars.

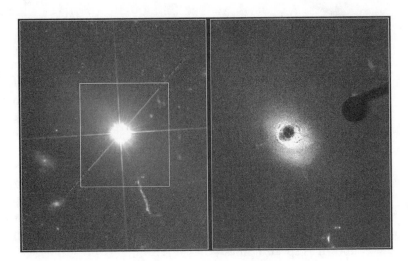

Quasar 3C 273. It looks pretty much like an ordinary star, but the faint jet that extends from it makes it only quasi-stellar. The jet shows clearly in the left image, from the Hubble Space Telescope. On the right image, taken with Hubble's new "coronagraph" ability, the bright central part of the quasar was blocked out, revealing structure in the quasar's host galaxy.

(Left image: NASA and J. Bahcall [IAS]; right image: NASA, A. Martel [JHU], H. Ford [JHU], M. Clampin [STScI], G. Hartig [STScI], G. Illingworth [UCO/Lick Observatory], the ACS Science Team and ESA)

Another key to understanding the spectral lines is the set of lines from ionized calcium. Although there are thousands of times fewer calcium atoms than hydrogen atoms, most of the absorption done by these calcium atoms takes place in the visible. By contrast, most of the absorption done by hydrogen takes place in the ultraviolet at wavelengths too short to come through the Earth's atmosphere. The converging series of hydrogen lines that we see are merely a secondary, weaker set.

In Fraunhofer's notation, one of the lines of ionized calcium is almost exactly the same color as the fifth line in the visible series of hydrogen's Balmer series, a line known as H-epsilon. This line has Fraunhofer's notation of H. But the H-line is only one of a pair; the other line of the pair, close in color in the part of the ultraviolet on the edge of visibility, has been given the notation K. The H- and K-lines are about equally strong.

Using mainly the Balmer series and the H- and K-lines, you can classify the spectra of stars. If you look only at the position where either H or the fifth hydrogen line would be, you can't necessarily tell which you are seeing. But by looking nearby at the K-line, you can tell how strong the H-line would be. By looking at the rest of the Balmer series, you can tell how strong H-epsilon would be.

In the early years of the twentieth century, a team of computers—humans who computed—worked on the spectra of stars at the Harvard College Observatory. The most famous was Annie Jump Cannon, who classified the spectra of hundreds of thousands of stars. The stars with the strongest set of hydrogen lines were called type A, the ones with the next strongest set of hydrogen lines were called type B, and so on.

It later turned out that hydrogen lines were strongest at a certain temperature and weaker at both hotter and cooler temperatures than that. When the series of spectral types was rearranged to be in descending order of temperature rather than descending order of strength of hydrogen lines, the order came out OBAFGKM. They were then subdivided in tenths, so part of the list would be A8 A9 F0 F1 F2 ….

Our Sun is a type G2 star. That classification comes from the fact that its H- and K-lines are relatively strong, while it still shows moderate hydrogen lines. Many other hundreds of lines are detectable easily, and tens of thousands of lines are visible if you look very carefully.

The hottest stars are O stars, with surface temperatures of about 60,000 kelvins. The Sun has a surface temperature of about 5,800 kelvins. For a long time, the coolest stars known have been M stars, at about 4,000 kelvins. Work around the turn of the twenty-first century, especially in the infrared, where the black-body curves for even cooler stars peak (see Chapter 8 for a discussion of black bodies), has led to the need for more spectral classes. Cooler than M stars are spectral type L. (L was the closest letter to M that wasn't already in use or confusible with something else.) M and L stars have many molecules showing in their spectra, in addition to spectral lines from atoms. Only in very cool stars can molecules survive. Some of the L stars, and an even cooler spectral type called T, are a type of failed star called brown dwarfs. Brown dwarfs did not get quite hot enough for hydrogen fusion to begin. They may have some fusion of deuterium in them, though.

We often keep track of types of stars by the masses they have. The stars with the highest masses are the intrinsically brightest and hottest. O stars, in particular, have about 60 times the mass of the Sun. M stars, on the other hand, have only about $\frac{1}{100}$ the mass of the Sun.

The Sun is thus typical of stars, in that it is in the middle of the range of stellar brightness. It is also in the middle of the range of stellar mass; there are cooler stars and hotter stars.

However, the Sun isn't just average. There are many more cool stars than hot stars. Thus, there are many more stars cooler than the Sun than there are hotter stars. There are many more less massive stars than there are more massive ones. Though its properties are average, the Sun itself is in the hottest, brightest, most massive group of stars.

The Least You Need to Know

- ◆ The Sun formed from a collapsing, spinning cloud of gas and dust.
- ◆ The Sun is mostly hydrogen, has a substantial amount of helium, and contains traces of all the other elements.
- ◆ We find the abundances of the various elements in stars by studying their spectra.
- ◆ The spectra of stars tell us their temperatures.
- ◆ The Sun is midway between extremes of stars in mass, brightness, and temperature.

Chapter 7

The Sun and Civilization

In This Chapter

- ◆ The Sun and Stonehenge
- ◆ Who found sunspots?
- ◆ Sun predictors
- ◆ The Sun as a cathedral
- ◆ The Maya and the Sun

It would be interesting to be able to travel back in time to ask people what they thought the universe was like. Books and other writings show us early views of the universe formulated over the last couple of thousand years. But what did people before that think? We have only the monuments and other things that they left behind.

Stonehenge as an Observatory

If you drive west and a little south from London, England, for a couple of hours, you spot some huge standing stones alongside the roadway on Salisbury Plain. You needn't stop, but you will be glad if you do. These stones are monuments from 4,000 or 5,000 years ago.

The stones are now protected by fences, although visitors used to be able to walk among them. In any case, they stand in a circle and two horseshoe shapes, with a few odd stones in various directions. In all, there are over a dozen stones twice as tall as a person. Each stone weighs about 25 tons.

Stonehenge, in England, with tall standing stones, some of which were oriented more than 4,000 years ago so that pairs were aligned to astronomically significant points.

(Jay M. Pasachoff)

The site is Stonehenge, one of Britain's National Heritage locations. It was built in several stages. The oldest stage, known as Stonehenge I, began about 3000 B.C.E. when holes in the ground were dug and wood posts were placed in them. Several hundred years later, about 2450 B.C.E., Stonehenge II was built, with the giant stones we see today brought in from distances as great as hundreds of miles.

The circle of stones is over 30 yards (27 meters) wide. Stone lintels link pairs of these giant stones. The horizontal links had to be raised 12 feet (4 meters) above the ground, probably a tiny bit at a time.

What were these giant stones? Why are they here? Why are they arranged as they are? Tradition holds that Druids prayed there, and modern-day Druids again meet there at certain times of year.

A key to understanding Stonehenge comes from the alignment of the stones. We have already seen that the Sun sometimes rises due east and, most of the time, rises north or south of due east. It rises due east at the days of the equinoxes. As we approach the summer solstice on June 21, the sunrise occurs as much as 23½° along the horizon north of due east. It can't be just a coincidence that the sightline from the center of the circle of stones toward an outlier stone almost 200 feet (60 meters) away points just to that direction. As was rediscovered in 1771, the Sun rises directly over this heelstone on June 21.

Or can it? There are a lot of stones and, therefore, a lot of directions mapped out by points of them. Archaeologists were skeptical when the astronomer Gerard Hawkins pointed out the coincidences between directions mapped out by pairs of stones and the directions of the Sun at key times of the year. But over the past decades, the idea has been widely adopted. Stonehenge is a type of observatory that marks key directions toward the Sun—not only the direction toward solstice sunrise, but also directions toward the equinoxes. Directions significant for the Moon also exist among the various alignments. The trick is to prove that these alignments are intentional, not just coincidental, given the many ways that paths can be drawn between any two stones in a large collection.

Why were the stones erected in that way? Why did the huge standing stones, known as sarsens, get transported hundreds of miles (hundreds of kilometers) at a time when such transport would have been incredibly difficult? Surely the people there at the time didn't make an observatory for scientific purposes to study the Sun.

Perhaps some religious ritual was involved. It is easy to see that an ancient people, not comprehending the motions of the solar system as we have since Copernicus and Kepler, might have placed great significance on the Sun's slowing of its northern travel from sunrise to sunrise. They may have wanted to celebrate the time when the Sun stopped traveling north from day to day and started traveling south. Indeed, during a few days around the solstice, the sunrise moves imperceptibly from day to day, a period that could have been a celebration. Perhaps the date was important for fixing a calendar that could be used for agricultural tasks, like planting.

One can readily imagine that people marked the direction of the rising sun on the horizon from day to day, perhaps digging holes or erecting small stones or pieces of wood to provide sighting directions. Such activity shows the existence of a stable society, able to stay in one location long enough to carry it out. Only after hundreds of years had passed and the available resources and technology had increased did the people decide to erect huge stones where once they had had flimsier markings.

A less widely accepted idea is that an outer ring of holes, known as Aubrey holes after their discovery by John Aubrey in the seventeenth century, acts as a predictor of solar and lunar

The Solar Scoop

People today can have fun mocking Stonehenge. In Alliance, Nebraska, a site called Carhenge uses cars standing on end and erected as lintels to mimic Stonehenge. Carhenge, like Stonehenge, has popular websites.

eclipses. The scheme advanced to use the Aubrey holes as eclipse predictors is complicated, and it is by no means clear whether the people of that time could have found it. Among the other difficulties is the fact that, most of the time, at that location in England, the sky is cloudy—even when an eclipse occurred, the people might commonly have missed it. (Even if the weather had been better overall at that time, many eclipses still would have been missed.) It would have taken decades of sightings before any pattern could have been established. The idea that Stonehenge is an eclipse predictor does not find general favor at the present. If it did, some say that scientific knowledge in England would have been more advanced than is currently thought. Accepting this chronology would force historians to change their ideas about which part of the world gained this knowledge first, and therefore about the direction in which knowledge diffused in the ancient world.

Meetings are still held to assess Stonehenge and what we know about it. From the facts on the ground, we try to imagine the minds of its creators. Even when we have a plausible understanding, that doesn't mean that it is accurate. We may be mirroring our own hopes and desires more than what the ancient peoples thought. Referring to the role of individual interpretation, it has been said, "Each generation gets the Stonehenge that it deserves."

Other "henges," circular arrangements of standing objects, have been located elsewhere in Britain. In recent years, a "woodhenge" has emerged in a tidal region, and archaeologists are trying to protect it. Many of these additional henges also have alignments between significant astronomical directions.

Where Comes the Sun?

Stonehenge isn't the oldest monument that has been linked to ancient astronomy. Giant mounds found in the Boyne Valley in Ireland contain huge tombs from more than 5,000 years ago. Newgrange, from 3500 B.C.E. to 2700 B.C.E., is a major stop on the tourist route.

The Solar Scoop

Too many tourists? Go early in the day to get a ticket to enter Newgrange, or the line may be too long for you to get inside.

Newgrange is thought to be an ancient burial place. Much of it has been excavated. A long tunnel leads from the outside into a central room whose ceiling is 50 feet (17 meters) high. On the day of the winter solstice, when the sunrise is as far south as it ever gets, sunlight passing through a slit over the entrance passes down the entire length of the tunnel for about 15 minutes. Surely this must have been on purpose.

The view from Newgrange and nearby shows many tombs. Among the excavated ones is a substantial tomb known as Knowth. Though tunnels at Knowth are not as long as the entrance one at Newgrange, they are also oriented toward directions significant for the positions of the Sun and the Moon.

The front of Newgrange, as reconstructed, in the Boyne Valley in Ireland. On the winter solstice, sunlight passes through a slit over the entrance shown and illuminates the length of the tunnel beyond.

(Jay M. Pasachoff)

In all these Boyne Valley Stone-Age sites, many of the stones have been carefully carved. Patterns such as whirls were incised. Clearly, a major effort was placed on making these sites ceremonially beautiful, beyond any attempt at mere astronomical significance.

Who Found Sunspots?

It is ordinarily hazardous to look at the Sun directly. Only during a total solar eclipse, when the everyday Sun is completely covered by the Moon, can you safely stare sunward. Usually, the eye's blink reflex prevents you from staring at the Sun, which is too bright. But occasionally the Sun is dimmed by haze to just the brightness at which you can safely look at it without any filter. This situation happens perhaps most often at sunrise or sunset, when the Sun is on the horizon, but it can happen even when the Sun is higher in the sky.

Over the centuries, a sunspot on the Sun occasionally was so large that it could be seen even with the naked eye looking through haze. So the true date of the first sighting of sunspots is impossible to pin down. Ancient Chinese, Korean, and Japanese records recorded sunspots some 2,000 ago.

With various reports of the first detections with a telescope, though, you might think that the date of the telescopic sunspot discovery might be accurately fixed. Here too, though, there is controversy and uncertainty.

Though Galileo may not have been the first to conceive of the idea of a telescope and to see a distant object with it, he was the first to use it successfully to study the heavens. In 1609, he turned his telescope upward and soon discovered the mountains and craters on the Moon, the phases of the planet Venus, the moons of Jupiter, and many other things. His book *Sidereus Nuncius*, published in 1610, contained all those discoveries. He went on to use the newfangled telescope to study the Sun in 1611.

Galileo, a portrait that appeared in his 1613 book on sunspots.

(Jay M. Pasachoff)

Though at first Galileo looked directly at the sun when it was faint enough at sunset, soon a student of his developed a technique of using the telescope to project an image of the Sun onto a screen. Then he could safely look at the screen. This method of "eyepiece projection" is still a popular and safe way to see sunspots. In his book on sunspots from 1613, Galileo drew the positions of the sunspots every clear day for a while.

Fun Sun Facts

Galileo studied both the night sky and the Sun with a telescope, and he had good vision up to the ripe old age of 72. So it wasn't true that he went blind because he looked too much at the Sun, given that blindness from Sun watching happens quickly. Galileo undoubtedly went blind from glaucoma and cataracts, which might have been treated had he not been prevented from seeking medical care during his house arrest from 1633 to his death in 1642.

But Galileo was not allowed a triumph over the discovery of sunspots. The Jesuit astronomer Christopher Scheiner also claimed priority. The rivalry between them for credit grew quite bitter and may have played a major role in Galileo's ultimate condemnation by the Roman Catholic Church. Scheiner at first had thought that the dark spots he saw were images of Mercury in silhouette, but he came to realize that they were really on the Sun. He wound up following sunspots carefully for 15 years and became the authoritative source on them. His masterwork on sunspots was published between 1626 and 1630.

So who first saw sunspots with a telescope? English scientist Thomas Harriot drew an image of one in late 1610 at about the same time that Galileo first saw them. David Fabricius and his son, Johannes, in Germany, and Scheiner first saw them in early 1611. Johannes Fabricius wrote about them in 1611, before either Galileo or Scheiner wrote letters about them in 1612, which were published in 1613 and 1612, respectively.

Fun Sun Facts

Even earlier than Galileo and Scheiner's fight over credit for discovering sunspots, Johannes Kepler saw a dark spot on the Sun in 1607 when he was trying to observe a transit of Mercury. He probably saw a sunspot instead.

 Solar Scribblings

David Fabricius was killed about five years after discovering sunspots when he accused a peasant of stealing a goose and, in return, was hit on the head with a shovel.

Cathedrals and Sunshine

Stonehenge was not the last of the ancient solar observatories. The Sun was considered a marker of significance in Europe from medieval times through the eighteenth-century Enlightenment period. John L. Heilbron, of the University of California, Berkeley, has shown that many of the most beautiful and significant cathedrals have holes in the roof and brass markers on the floor to follow the changing position of a solar beam of light over the year. Some of this astronomical application was fashioned in the seventeenth and eighteenth centuries.

The Roman Catholic Church has long been interested in calendars, especially in order to fix the date on which to celebrate Easter (lest people's souls suffer for celebrating Easter on the wrong day). In order to fix the date properly, the Church needed the best observations possible. So quite aside from its qualms about theories of the universe, the Church methodically, if quietly, saw that excellent data were collected. In particular, the Church used the meridian lines, the north-south brass lines put on the floor of many cathedrals for that purpose, to monitor exactly when the Sun returned to its yearly position in the sky at the time of the vernal equinox, a time not far from Easter. Calculating the date of Easter depended on this interval and on the intervals between full moons. Easter was set for the first Sunday after the first full Moon after the vernal equinox.

> **Fun Sun Facts**
>
> On May 28 and June 6 each year, the rising and setting Sun shines straight down each street in Manhattan's grid.

In 1543, Nicolaus Copernicus advanced his theory that the Sun, not the Earth, is the center of the universe (we now limit that centrality to the solar system), and the Church worried that his theory might contradict the Bible. Still, Copernican theory allowed astronomical events to be predicted more accurately than the earlier theories, so the Church used it all the time. However, as a fiction, the Church supposedly used it merely as a basis for calculation. Indeed, a clergyman (though not a Roman Catholic one) added the phrase "of the celestial spheres" to Copernicus's original title, *On the Revolutions*, to indicate that the work might apply only to calculations rather than to truth. Only with Copernicus's theory based on the Sun at the center could the date of the vernal equinox be predicted.

Among the cathedrals that were used as solar observatories were those in Bologna, Rome, Florence, and Paris. Huge obelisks outside the cathedrals carried sundials. Another meridian is at the Tower of the Winds in the Vatican. The data were used in advancing the Gregorian calendar, promulgated in 1582 by Pope Gregory XIII, which is the common calendar in use today.

Heilbron concludes that the Church's great interest in astronomy disproves the widely accepted notion that it was hostile to science, something that was most emphasized in its battle with Galileo. As long as clerics were able to say that Copernicanism was a theory rather than fact, they could teach any astronomy they liked. Indeed, Owen Gingerich of the Harvard-Smithsonian Center for Astrophysics has examined the 500 extant copies of the first and second editions of Copernicus's book. He found that only a few of them, and chiefly those in Italy, were censored (with supposedly offending passages blacked out), while the others circulated in their original form.

Even today, moving sunlight can be used in memorials. The winning proposal in 2003 for rebuilding the World Trade Center site in New York has a wedge of sunlight, defined by the new buildings, falling on the memorial each September 11 during exactly the hours that the twin towers stood after being hit on 9/11/01. The idea is reminiscent of solar markers in cathedrals of old.

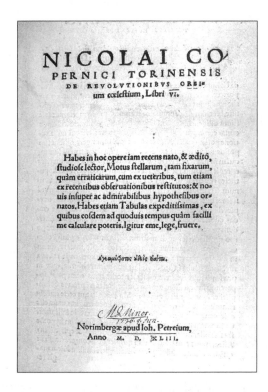

The title page of one of the fewer than 300 extant copies of the first edition of Copernicus's De Revolutionibus. This copy of the book is bound in leather, typical of northern European rather than Italian bindings. Its northern-European association is consistent with the idea that the pages have not been censored, something that occurred more frequently in Italy.

(Jay M. Pasachoff)

Fun Sun Facts

The great thirteenth-century painter Giotto is famous for his set of frescos in the Arena Chapel in Padua, Italy. On one frame in the sequence, a model of the chapel is shown, with the donor of the chapel offering it to the Virgin Mary. Each March 25, the feast of the Assumption and the date on which the chapel was consecrated, a ray of sunlight falls on the model. Before a door was closed permanently, another ray of sunlight through that door lighted up Christ's halo in the *Last Judgment*.

South of the Border

No sites in North America are older than those in Britain, but we have both monuments and written materials from the Maya. The Maya flourished in what is now Mexico and Guatemala about a thousand years ago. Among the huge structures they left behind are giant pyramids at sites like Chichen Itza in Mexico's Yucatan peninsula, Tikal in Guatemala, and the Pyramids of the Sun and of the Moon outside Mexico City. Archaeoastronomers—people who study both archaeology and astronomy— have measured various positions and angles accurately at these sites and have found many alignments to significant astronomical directions.

These sites are recent enough that we have hope of finding out about the intentions of the creators from written material. The Aztecs made many handwritten books, bound together in fold-out volumes called codices. Unfortunately, when the conquistadors came from Spain, they wanted to destroy and overwhelm the civilizations they found in America. They burned all the codices they could find. Only a handful survived. One is in Madrid, Spain (now called, not surprisingly, the Madrid Codex); another notable one, the Dresden Codex, is in Dresden, Germany. The Madrid Codex once was on display at the Museum of the Americas on the university grounds in Madrid, but now it is kept in a vault and only a high-quality facsimile is on view.

Some of the symbols in the codices are thought to represent the Sun, the Moon, Venus, and eclipses. Unfortunately, we aren't always sure which. Still, the numbers have been decoded, and it is known that the Maya had an accurate calendar that was based not only on the Sun and the Moon, but also on the rising and setting of Venus. The Maya obviously followed Venus's position over many years, a sign of a stable civilization, in order to be able to track how often it rose as the morning star just ahead of the Sun. From this information, known as Venus's heliacal rising, they discovered that Venus had a periodic cycle of visibility. The Mayan calendar was more accurate than European calendars in use at that time.

It is still debated why Mayan civilization failed. Perhaps drought or warfare caused the collapse of their complex interactions. We suspect there is a lesson for our own civilization, even though we don't know what it is.

The Maya were not the only American civilization with astronomical alignments. Substantial constructions at Monte Alban and Mitla near Oaxaca, farther west in Mexico, are also of interest for archaeoastronomers because of solar alignments.

Astronomy of "The People"

In the thirteenth century, the Anasazi tribe (the world *Anasazi* means simply "The People" in their language) built elaborate dwellings in what is now Arizona and other parts of the American west. Some of the dwellings show astronomical orientations. For example, the setting sun is visible straight through a particular window only on mid-summer night.

One of the best known phenomena is the "sun dagger" found in Chaco Canyon. A spiral was incised on a large rock. On a certain day of the year, a narrow strip of sunlight, like a dagger, descends on the rock after passing through a space between other rocks. It comes right through the middle of the spiral.

The Solar Scoop

Many people have visited the sun dagger and other sites in Chaco Canyon in recent years. Unfortunately, the rocks near the sun dagger have now settled, so the archaeoastronomical phenomenon no longer can be viewed.

Archaeoastronomers such as Edward Krupp, Jack Eddy, and Anthony Aveni continue to study various American sites. A giant circle of stones built between 1400 and 1700, for example, has attracted their attention. This Bighorn Medicine Wheel (here, *medicine* has the meaning of "magic") contains various alignments toward significant astronomical points. In particular, some point to the directions of sunrise and sunset on the day of the summer solstice. Others point to the rising points for bright stars at significant times of year. Krupp concludes from these and other alignments, and from his knowledge of Cheyenne rituals, that the astronomical alignments were put there on purpose. It is at least plausible that they were part of some rituals.

About 50 medicine wheels are known, ranging in size up to hundreds of meters (hundreds of yards) across. Some are thousands of years old, ranging back to the time of Stonehenge in Britain. Many of these other medicine wheels also have solar alignments.

The Least You Need to Know

- Stonehenge demonstrates knowledge of the Sun's motion from 5,000 years ago.

- Boyne Valley tombs show alignment to significant solar directions even earlier.

- Sunspots were first observed through a telescope by Galileo or perhaps by Scheiner.

◆ Some medieval cathedrals marked the changing direction toward the Sun.

◆ The Mayan calendar was based on the Sun, the Moon, and Venus.

◆ Native American sites from hundreds of years ago had solar alignments.

Chapter 8

The Birth of the Sun

In This Chapter

- Young stars glowing like toasters
- H-H objects sending off jets
- T Tauri stars flickering
- The Sun is a dwarf
- How old are you, Mr. Star?

We love the Sun because it is ours, but, in the words of Gilbert and Sullivan's comic operetta *H.M.S. Pinafore*, we also need to study "his sisters and his cousins, whom he reckons up by dozens, and his aunts." Studying many stars, singly and in groups, tells us how stars like the Sun formed and shows us similar objects today.

Flying Toasters in the Sky

Toasters are known to fly across computers as screensavers and traditionally have been given out by banks for opening accounts, but did you know that they can remind us of young stars? If you look inside a toaster as you turn it on, you can see that the wires soon begin to glow dully. Then they begin to turn a little reddish. Later they glow bright red. We are seeing what astronomers call black-body curves.

A black body is a perfectly radiating body. By "black," we mean that it isn't polka-dotted or otherwise varying from place to place. It is uniform, and we can describe the radiation that it gives off by only one number. We call that number the temperature. The temperature really describes how fast the particles of matter are moving every which way.

The Solar Scoop

The peak of the Sun's spectrum is in the yellow-green part of the spectrum, so we humans may well have evolved to have our greatest sensitivity there. Since Superman has x-ray vision and was born on the planet Krypton, can we conclude that Krypton's star is much hotter than the Sun and thus shines more brightly in x-rays?

A black body gives off radiation, but how much radiation it gives off and the amounts of radiation at different colors depend on its temperature. For any radiating black body, there is a certain color at which it gives off more energy than any other color. At colors corresponding to longer wavelengths, it gives off less energy. At colors corresponding to shorter wavelengths, it gives off less energy. If we graph the energy that it gives off as a function of wavelength, we see a peaked curve (see the following figure). Sometimes this curve is called a black-body curve. It is also often known as a Planck curve, after the scientist Max Planck, who found a mathematical formula for it just over 100 years ago.

A toaster heat element is about 1,000°C and thus glows red hot. At that temperature, though most of the radiation is in the infrared, a small fraction is in the visible part of the spectrum, so we see the toaster wires begin to glow visibly. The hotter the wires become, the larger the fraction of energy that is emitted in the visible.

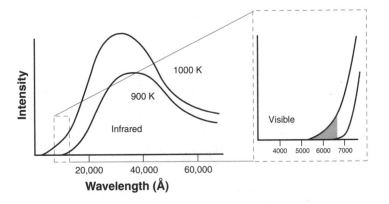

When you heat up an object, the peak of its radiation starts way out in the infrared. As you heat it up more, the peak moves to shorter and shorter infrared wavelengths—that is, it moves closer to the red. At this temperature, you see the toaster (or a star being born) glowing reddish. As it heats up more, it gives off more energy in the red, though the peak of the radiation is still out in the infrared. When it gets hot enough, to several thousand degrees, the peak of the radiation actually moves into the red. Then, when it reaches the 5,800 °C of the Sun, the peak has actually moved beyond the red into the yellow-green.

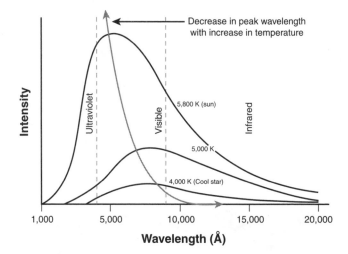

Wien's Displacement Law tells us where the peak of radiation for a black body is. An object at 4,000 kelvins peaks in the red, while at object at 5,800 kelvins peaks in the yellow-green and an object at 7,000 kelvins peaks nearly in the blue.

Since the Sun gives off most of its radiation in the yellow-green, we humans have evolved so that our eyes are most sensitive there. Our eyes have rods that sense even dim light and cones that come in three types, each sensitive to a different color. Used together, the cones give us our perception of color.

As a cloud of gas contracts, it first gains its energy from gravity. That energy is enough to make it glow slightly. But it has to turn on and start nuclear fusion to get enough energy to become yellow-hot. Fortunately, the toaster in your kitchen never gets that hot. But an iron poker in a fire can go beyond red-hot to yellow-hot and even to white-hot.

Fun Sun Facts

Sugars and starches caramelize, making toast, at about 320°F (160°C). When they carbonize, at still higher temperatures, you've burned your toast.

Riding Madly Off in All Directions

In *Gertrude the Governess, a Nonsense Novel* by Canadian humorist Stephen Leacock, a hero "flung himself upon his horse, and rode madly off in all directions." Something similar happens with young stars. Since they are spinning, there is a preferential direction along their poles for gas to escape. With the aid of the magnetic field, gas is squirted out along the axes into space.

This gas was seen before it was understood. American astronomer George Herbig and Mexican astronomer Guillermo Haro found some dozens of places in the sky where bits of nebulosity seemed to be associated with young stars. They cataloged them; they are now called Herbig-Haro objects and often called simply *H-H objects*. Modern telescopes, now including the Hubble Space Telescope, were required to see them clearly. These newer observations show that the gas is ejected in opposite directions from a young star (see the following figure). In the images, sometimes the star itself is buried in dust so that we can't see it. In some of these cases, looking with telescopes and detectors sensitive to the infrared, we get radiation that penetrates the dust, and we can see through to the central star. With these optical and infrared observations, H-H objects have been transformed from mere curiosities to important signals of how stars like the Sun form.

Sun Words

H-H objects are jets of gas given off by young stars.

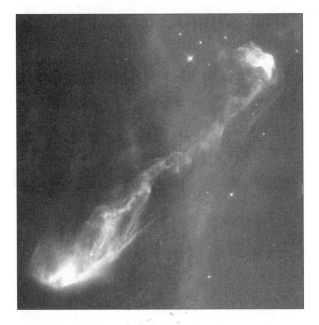

A Hubble Space Telescope view of HH-47. We see gas being ejected by a young star, though the star itself is hidden by dark clouds of gas. This view reveals that the jet has punched a cavity through the dense gas cloud into interstellar space. The wobble in the jets suggests that the hidden star might be wobbling, possibly because of the gravity of a companion star.

(J. Morse/STScI, and NASA)

Flickering Starlight

When Copernicus's famous book *On the Revolutions* was published in 1543, modern astronomy began. Copernicus put the Sun instead of the Earth at the center of the universe so that we live in what we now call the solar system. But Copernicus still had fixed stars around the edges of his diagram of the solar system. It was a big surprise when, only about 50 years later, in 1596, David Fabricius discovered a star that varied in brightness. He called it *Mira*, from the Latin for "miraculous." We still call the star Mira, though it also has its place as omicron Ceti in the labeling scheme set out by Johann Bayer in his beautiful star atlas of 1603.

A couple hundred years ago, the deaf-mute amateur astronomer John Goodricke went on to discover several other important variable stars, including the Cepheid variables that allow astronomers to tell the distances to other galaxies. These variable stars are at the center of our understanding of how old the universe is and are, therefore, also at the center of our understanding that a Big Bang occurred some 14 billion years ago.

Sun Words

Mira was the first variable star discovered.

The Solar Scoop

A wonderful amateur astronomy association, the American Association of Variable Star Observers (AAVSO), collects and assembles observations of variable stars made by people around the world. On the association's website, you can get the light curve—the graph of brightness over time—of any variable star for any period of time. You can also get a finding chart to show you how to find the variable stars of your interest in the sky. See www.aavso.org.

The first variable star in a constellation is known as R with a form of the name of that constellation. Thus the first variable star in the constellation Corona Borealis is known as R Coronae Borealis, or R CrB. The next variable star is S, the next is T, and so on. After Z, you start with RR, as in the important variable star RR Lyrae. Now we know of hundreds of variable stars in each constellation, so we are merely numbering them; the letter designations have long since been exhausted.

Thus, from the name *T Tauri*, we know that this star was the third variable star discovered in the constellation Taurus, and it must have been known as variable for a long time. T Tauri itself is just one of many similar objects known as T Tauri stars. These are stars that are now in the stage that the Sun was in about five billion years ago.

Sun Words

T Tauri stars are young stars that are still unsteady in brightness, resembling the early years of the Sun's life.

T Tauri stars vary irregularly in brightness. When astronomers take their spectra, the spectra show emission spectral lines, revealing that there is still hot gas around the star that hasn't disbursed or collapsed into the star's surface.

The Sun was a T Tauri star about five billion years ago, but our star has grown out of this adolescence. T Tauri itself has now been studied in various parts of the spectrum. Images in the infrared have revealed that it is actually a double star with a nearby companion, which makes it different from our Sun.

The Least You Need to Know

- The sun was first red-hot and now is yellow-hot.
- Newly formed stars that resemble the early Sun give off beautiful jets of gas.
- Cepheid variable stars allow us to find the distances to galaxies and the age of the universe.
- Variable stars like T Tauri haven't yet settled down to radiate steadily.

Chapter 9

The Sun at the Center

In This Chapter

◆ Copernicus put the Sun at the center

◆ Galileo and Kepler figured things out

◆ The Sun isn't perfect

◆ Other solar systems

Throw Down Your Ptolemy

More than 2,000 years ago, about 350 B.C.E., the great philosopher and scientist Aristotle advanced a theory of the universe that seemed to make sense, at least to our ordinary senses: We are on a stationary Earth, and the Sun, the Moon, and the planets go around us. About 500 years later, Claudius Ptolemy, who flourished in Alexandria around 140 C.E., modified Aristotle's theory to make it fit better with the details of the observations of the time. In particular, he had to account for the details of the wanderings of the planets in the sky with respect to the stars.

If you watch the sky every night, you can see bright planets like Jupiter or Mars rise and set with the stars. If you watch from night to night, though, you can see that they appear to drift among the stars. For most of the time, they move slightly faster than the stars in their rising and setting; we call such motion prograde. And occasionally, they seem to fall behind the stars in their rising and setting; we call such backward motion retrograde.

Ptolemy tried to reason how the planets could sometimes appear to move backward. He suggested that rather than merely being attached to huge spheres that rotated around Earth, they were on small circles whose centers moved on the big spheres. The main motion on the spheres as they turned were said to be on the deferents. But motion on the small circles, known as epicycles, brought the planets apparently backward for part of the time.

In the early 1500s, Polish canon, physician, and astronomer Nicolaus Copernicus worked out a system of the universe in which the Sun instead of Earth is at the center. He explained retrograde motion as the effect of perspective. As an example, you have been in a car at a stoplight and thought you were going backward as the car next to you drifted forward. The effect in the solar system is the same: We might see Mars seem to drift backward as we pass it as we both orbit the Sun.

The first heliocentric diagram, from Copernicus's book of 1543. Note that the Sun (sol, in Latin) is at the center.

(Jay M. Pasachoff)

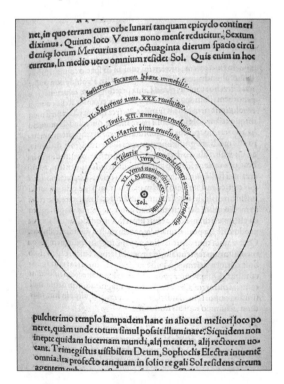

We have accepted for years that Copernicus's work demoted the Earth from the center of the universe to a peripheral role. Recently, Seattle English professor Dennis

Danielson has realized that a careful historical reading shows that the center of the system was not a desirable place to be. After all, sewage winds up at the center of a system on Earth, and, to Aristotle, "gross matter" settled at a the center of the world. We may be well out of that filthy place.

Solar Scribblings

Anthony Hecht and John Hollander invented a verse form known as the double dactyl, an eight-line poem in which every line has the "da' da da" dactylic form repeated twice. The first line is a nonsense set of syllables, and one line must be a single double-dactylic word. One of their efforts comes to mind here:

Higgledy-piggledy
Nic'laus* Copernicus
Looked at the universe,
Spoke to the throng:

Give up your Ptolemy,
Rise up and follow me,
Heliocentrically
Ptolemy's wrong.

*This is clearly cheating; on the other hand, "Nicky" would have been far, far worse. Lines 5 and 6 are nevertheless inspired.

(Poem by Nancy L. Stark; footnote by John Hollander. From *Jiggery-Pokery*, Copyright 1966, 1983 by Anthony Hecht and John Hollander. Used by permission.)

Galileo on Your Side, Nicky

One of the most brilliant scientists at the turn of the sixteenth century, when Shakespeare was treading the boards and writing his plays, was Galileo Galilei, in what we now call Italy. As we have seen, he turned the newfangled telescope at the sky in 1609. He had come to believe that Copernicus was right, that the Sun was at the center of the solar system. Among his early achievements, he had figured out the precursor of what we now know as the law of inertia, and with it he could understand why some of the supposed arguments against Copernicanism were wrong. For example, Copernicus's opponents asked how birds weren't whisked away by huge winds if the Earth were spinning rapidly in space. Galileo realized that the birds and the air shared in the Earth's rotation and kept going by inertia.

Galileo made his early discoveries in physics while at the University of Pisa. For example, he discovered that a pendulum swings with a constant period no matter how high it swings. Then he moved to the University of Padua, which enjoyed the freedom of the Venetian Republic. He was such a good lecturer that a tall platform and chair were built for him to stand on and sit in to allow students to see him while he lectured. These objects still survive at the University of Padua. He was at that university when he worked on his first telescopes.

Galileo's telescopic discoveries got him what he thought was a better job: back in Tuscany, in Florence. But Florence was closer to Rome and was under the influence of the Pope and the Inquisition, which was rooting out supposed heresy. In 1616, Galileo was brought before the Inquisition and was even shown the rack and other instruments of torture in order to convince him to withdraw his backing for the heliocentric theory. Rumor has it that as he left the room after abjuring, he muttered "And still it [Earth] moves."

Galileo kept himself clean until 1629, when he couldn't resist anymore. An old friend was then pope. Galileo wrote a book in the form of a discussion among three individuals discussing the theories of how the universe is arranged: *Dialogue Concerning the Two Great World Systems*. It took him three years to arrange permission to publish it, and it came out in 1632. The book was written in the vernacular, common Italian that the general public could understand, instead of the Latin of scholarly works. That fact helped condemn Galileo, especially when the pope realized that at the end of the book, Galileo had placed a foolish argument in the mouth of a character identifiable as the pope.

Galileo was again called before the Inquisition. This time, he didn't get off. He was convicted of disobedience and sentenced to house arrest. Even while restrained, though, he continued to work and wrote important books, including one on optics. He died on Christmas in 1642.

Fun Sun Facts

The story of Galileo and his fight with the Roman Catholic Church was so dramatic that it is still written and sung about. Bertholt Brecht's play *The Life of Galileo*, dating from 1943, is an old standard. In 2002, the opera composer Philip Glass and the librettist and director Mary Zimmermann collaborated with librettist Arnold Weinstein on the opera *Galileo Galilei*, which played in Chicago, New York, and London. It used the theme of Galileo's eventual blindness, ironic in one who saw so far for the first time, to trace out his life, and it included dramatic scenes of Galileo with the Inquisition.

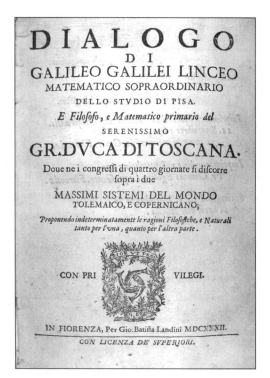

The title page of Galileo's Dialogo, *the book that got him in his final trouble with the Inquisition when it was published in 1632.*

(Jay M. Pasachoff)

Kepler Controls the Universe

Johannes Kepler was a young mathematician in Germany in 1601 when he wrote to the great astronomer Tycho Brahe, who had just moved to Prague. Tycho had brought with him his meticulously collected data about the positions of the planets, principally Mars, in the sky. Though the telescope had not yet been invented, Tycho had used large sighting instruments to record data of unprecedented accuracy.

Fun Sun Facts

Kepler was one of the greatest mathematicians of all times, and many of his achievements are still in use today. For example, Kepler figured out in 1611 that the way your oranges are displayed on a fruit stand or at the supermarket—with a base row, a second row filling in, and so on—is the most efficient way to pack spheres. When his supposition was finally proved in 1998, the achievement merited headlines. When Kepler got married in 1613, he noticed at the celebration that there was then no efficient way to measure the contents of the wine barrels. Over the next couple of years, he figured out a way to measure the contents of a volume enclosed by rotating a curved shape.

In 1600, Kepler joined Tycho's entourage in Prague. It included a household of dozens, including even a jester. When Tycho died the next year, Kepler wanted to continue work on the data, but the data were among the most valuable property of Tycho's heirs. Eventually, Kepler got his hands on the data and went to work on the orbit of Mars.

It took him years to figure out how Mars's orbit around the Sun led to the positions of Mars recorded in the sky, including the retrograde loops. Remember that there were no calculators or computers; he had to do all his work by hand. Indeed, almost two decades later, when logarithms were first invented, Kepler was the first to adopt them for scientific calculation. He was as up to date in his methods as possible, but he still had to carry out lengthy calculations in an arduous fashion.

In Kepler's book *Astronomia Nova* (*The New Astronomy*) of 1609, he announced his breakthroughs: The planets didn't orbit the Sun in circles. Rather, they orbited in a type of squashed circle known as an ellipse. The orbits weren't far out of round, but enough so that the correction gave much more accurate predictions for the positions of Mars. Later, clear statements were extracted of what we now call Kepler's Laws of Planetary Motion. Kepler's first law is this: The planets orbit the Sun in ellipses with the Sun at one focus. To draw an ellipse, hold a piece of string with its ends at two different points (the two foci, the plural of focus). Use a pencil to hold the string taut, and move the pencil around. The result is an ellipse. When the two foci are at the same place, the ellipse is the special case that we know as a circle. When the two foci are far apart, the ellipse is very far out of round and matches the orbits of many comets.

Kepler's book of 1609 contains his second law as well: "The line joining the Sun and a planet sweeps out equal areas in equal times." Thus, the long, skinny triangle swept out in a day must have the same area as a short, fat triangle covered in a day at another time of the orbit. As a result, when the planet is relatively close to the Sun, it must move faster in its orbit.

Kepler's third law, which links the period of the orbit (the time it takes to complete it) and the average size of the orbit, didn't come out until his book *Harmony of the World* was published in 1618.

Only with Kepler's laws did we have a clear view of the place of the Sun in the solar system. Kepler's laws are applicable even today in a wide variety of situations. For example, in 2002, observations were reported showing a star that orbits the center of our galaxy in only about 15 years. Given the size of its orbit, which we can measure from Earth only because of advances in optical telescopes, astronomers calculated from Kepler's laws that a giant black hole containing 2.5 million times the mass of the Sun is sitting in our galaxy's center.

The statue in Prague of Johannes Kepler and Tycho Brahe, with the author.

(Deborah Pasachoff)

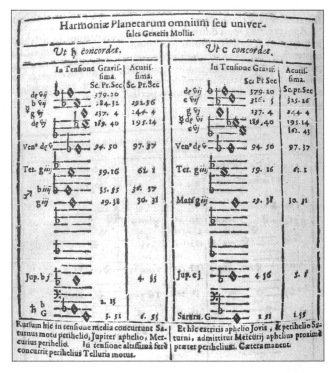

An excerpt from one of the pages of Kepler's book of 1618, showing the notes to illustrate the harmonies that he thought existed in the orbits of the planets.

(Jay M. Pasachoff)

Imperfection

Traditionally, under the ideas of Aristotle and Ptolemy, the universe was a perfect place, at least in the realm of the planets and the stars. In Aristotle's model, the planets and stars orbited on crystalline spheres. The orbits of objects, originally around the Earth, were thought to be perfect circles, and circular motion was thought to be natural. Even Copernicus used circular orbits, although his orbits were around the Sun rather than the Earth.

But Kepler's discovery, announced in 1609, that orbits weren't circular destroyed the idea of circular perfection governing the universe. About a year later, Galileo's work with the new telescope revealed that the surface of the Moon was pockmarked, so it also wasn't perfect. Then, about another year later, the realization that the Sun had sunspots on it showed that its surface wasn't perfect, either. We had to come to terms with a universe that was imperfect—blemishes existed even in the celestial realm. It won't be long before we reach the four-hundredth anniversary of that heady time of scientific discovery.

Dozens and Dozens of Planets

Until 1781, people knew of six planets: Mercury, Venus, Earth, Mars, Jupiter, and Saturn. People were stunned that year when in Bath, England, Hanoverian musician William Herschel discovered the planet soon to be named Uranus. In the 1840s, the world was further excited by the discovery of the planet that was named Neptune. National rivalries came into play. People in England still tend to give credit for the discovery to John Adams, who first predicted its position, while people in France credit Urbain Le Verrier, who soon thereafter predicted its position independently and published the results first, which led to the actual sighting. People in Germany credit Johann Galle, who first actually sighted this eighth planet.

The ninth planet, at least from 1930 to the present, is Pluto. It was discovered by Clyde Tombaugh, then a young astronomer, so until his death in 1997, we were privileged to live at a time when the discoverer of a major planet was still alive. Nowadays, our new, more powerful telescopes are discovering objects not too far from Pluto in size among and beyond similar objects, so perhaps Pluto isn't a major planet after all. It hasn't been officially demoted by the International Astronomical Union, and even the Division of Planetary Sciences of the American Astronomical Society has defended its planethood. But that's all another story.

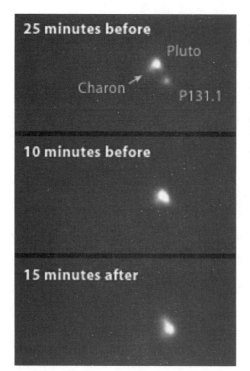

Pluto, with its moon Charon, about to occult (hide) a star in 2002. I observed the occultation in order to study Pluto's atmosphere, and this image was taken by my team at the University of Hawaii's 2.2-m telescope on Mauna Kea.

(Jay M. Pasachoff)

During the 1990s, a number of planets were found around sunlike stars—that is, distant, ordinary stars that have about the same amount of mass and thus the same brightness as the Sun. Many people are especially excited about planets around such stars because, with our limited knowledge of how life evolves, they seem to have the highest chance of having intelligent, communicating life on them.

The new planets were found by observing the spectral lines from the star and how they shift back and forth over time. The change in wavelength of light (or of any other wave) is known as the Doppler effect, after nineteenth-century physicist Christian Doppler. To make the discoveries, the scientists needed to develop new methods of measuring the Doppler effects in stars to a new level of precision.

The planets can be discovered because each system of objects has a center of mass that follows laws of motion set down in 1687 by Isaac Newton. First, you must know that all the mass of a spherical object acts as though it is at the center of that object. Proving that point took Newton years, and he had to invent calculus to do so. If in a pair of objects, one object is much heavier than the other, the center of mass is closer to the center of that object. So if you consider in our solar system only Jupiter and the Sun, both orbit the center of mass of the pair, but that point is located deep inside the Sun. So the Sun would appear to move back and forth only a little each time Jupiter goes around, while Jupiter swings from 5 A.U. out from the Sun (that is, 5 astronomical units, or 5 times the Earth's average distance from the Sun) around to 5 A.U. out the other side over about a 12-year period (that is, a period of 12 Earth years).

As we discussed in Chapter 1, the first of these planets around sunlike stars was found in 1995 by Swiss astronomer Michel Mayor and his graduate student Dedier Queloz. Their find was soon confirmed by American astronomers Geoff Marcy and Paul Butler, who had collected good data that showed the effect but who hadn't yet analyzed it. After all, since Jupiter takes about 12 years to go around the Sun, who was expecting a planet that went around a sunlike star in only 4 days?

With the 1995 discovery, the floodgates opened, and occasionally teams of scientists have listed as many as a dozen newly discovered planets at a time. Mostly these are individual planets around a given star, but, of course, we are discovering the most massive ones first. As our techniques improve, we will discover less massive ones, though it may be over a decade before our sensitivity improves enough to discover a wimpy planet like Earth. NASA satellites, including the Space Interferometry Mission (SIM) in this decade and the Terrestrial Planet Finder in the next, hold out hope. NASA's Kepler mission and the European Space Agency's Eddington mission, to detect planets transiting stars by a slight dip in a star's brightness, may join SIM in space in the later years of this decade.

> **Fun Sun Facts**
>
> Giordano Bruno was burned at the stake in Rome in 1600 for his heretical belief that there was a multiplicity of worlds, as we mentioned in Chapter 5. It took almost 400 years, but by that anniversary of his death, about 100 worlds were known to be orbiting stars other than the Sun. So Bruno was right, after all.

> **Fun Sun Facts**
>
> Who wouldn't want to get a Nobel Prize? But the rules say that no more than three people at a time can share it. How would you pick three for a prize on new planets? Queloz—who worked with Mayor—was only a student but was there first in the discovery of planets around Sunlike stars, while Marcy and Butler (since joined by Debra Fischer) found more planets than anyone. And planets around a pulsar were discovered even earlier. Somebody deserving will get shut out if and when a prize is given on this topic.

You can always get a list of planets around other stars at www.exoplanets.org, the site that Marcy keeps current. As of this writing, it shows over 100 planets.

The Least You Need to Know

- Copernicus's heliocentric theory eventually supplanted Earth-centered theories.

- Galileo championed Copernicus's theory in the early 1600s.

- Between 1601 and 1618, Kepler figured out the basic laws of orbits.

- Celestial objects like the Sun are not perfect.

- Astronomers are detecting solar systems around stars other than the Sun.

Chapter 10

The Death of the Sun

In This Chapter

♦ The Sun will swell

♦ The Sun will be a beautiful nebula

♦ The Sun won't be a supernova

♦ The Sun will be a cinder

We live our lives on a planet orbiting a middle-aged star. The Sun has been shining for about five billion years, and we have about another five billion years to go. From watching other stars more advanced than ours, we can tell what will happen. It's like watching our future.

Red Giants and Yellow Dwarfs

About 90 years ago, two scientists independently plotted measures of temperature and brightness for a bunch of stars on a graph. One, Henry Norris Russell of Princeton University, corrected his values for brightness to take into account how far the individual stars were from us. After all, something can look bright either because it is really bright or because it is merely very close. The other, Ejnar Hertzsprung, a Danish astronomer working in Germany, examined a lot of stars in star clusters, taking advantage of the fact that all those stars in each cluster were just about at the same distance from us as the others. Their result has dominated stellar astronomy ever since.

In particular, they discovered that stars weren't scattered all over their graph, which might have been the case. Instead, most stars formed a "main sequence," a fairly straight line diagonally across the graph. The horizontal axis of the graph in the following figure shows the stars' surface temperatures, which can be measured in various ways.

The Hertzsprung-Russell diagram, the plot of some measure of stars' temperature vs. some measure of their intrinsic brightness.

(The Hipparcos Project and ESA/M.A.C. Perryman)

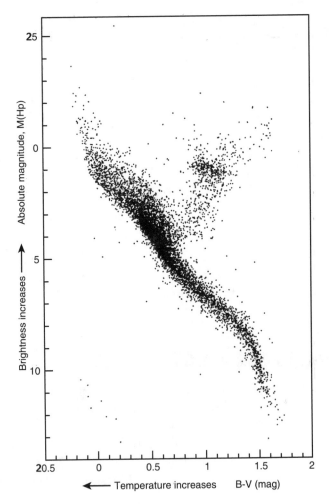

If you draw a vertical line on the preceding graph showing a color-magnitude diagram, that line hits a lot of points, each of which represents a star. Going upward from the bottom axis (that is, going to brighter stars), the line hits the curved band that represents the main sequence at some point. Most of the stars that the line hits, especially when you draw that vertical line for some red color, are on the main sequence. But a few of the points that are hit by the line are above the main sequence on the graph. That is, the stars are brighter than main-sequence stars at the same color.

The Solar Scoop

The diagrams that have descended from the work of Hertzsprung and Russell are known as H-R diagrams or color-magnitude diagrams, since color is a measure of temperature, and magnitude is a measure of how bright a star is. Harking back to the spectral types of stars, which are given the order OBAFGKM, the O stars, the hotter stars, are at the left. Therefore, temperature increases going leftward, which is opposite the usual sense of increase for that axis in most graphs. Also, the fainter a star, the higher its magnitude. When first called that thousands of years ago, the brightest stars were "of the first magnitude." So the magnitude scale on the vertical axis increases downward. In some sense, that is backward, too. So both axes, for historical reasons, are opposite of how most graphs are drawn.

In Chapter 8, we discussed black-body curves. Whenever gas has an overall smooth spectrum with a peak of intensity at some particular color (such as that shown in the first figure in Chapter 8), we say that the gas is acting like a black body and that all that gas is at the same temperature. Let us use this concept to compare two stars on the same vertical line (that is, of the same temperature), with one star higher (brighter) than the other on the graph. So both stars—the fainter one on the main sequence and the brighter one above it—have surfaces at the same temperature. That means that each bit of surface (let's say, each square inch) gives off the same amount of energy. So how can one star be brighter? It must be larger, since then it has more surface area. That makes the star a giant. By looking at how diffuse their spectral lines were, Hertzsprung had already noticed that these stars were very large (since larger stars have lower gravity at their surfaces, which allows the atoms there to move around faster, making their spectral lines less sharp).

In comparison with the *giant* stars, the stars on the main sequence are known as *dwarfs*.

Sun Words

Giant stars are bloated; their surfaces are larger than those of dwarf stars of the same temperature, giving them more surface area and making them brighter. **Dwarf stars** are normal stars; they are in the prime of life, not yet having gone through the death throes that transform stars to giants, supergiants, or even later stages of stellar evolution.

Though the name—chosen by Hertzsprung—may be odd, dwarf stars are normal stars, and they include the Sun. Hertzsprung just noticed that at a given color, some stars were smaller than the giants he had already studied, so he called them dwarfs.

In about five billion years, the Sun will have used up much of the hydrogen in its core, which takes up about 10 percent of its size. The hydrogen will be transformed into helium, and hydrogen will continue to fuse to helium in a shell around the core. But a stable star is in a constant battle between the force of gravity pulling everything inward and some pressure pushing it outward. When the pressure of hydrogen fusion in the core disappears, the star collapses a bit. The energy released from the collapse heats it up, and the core gives off more energy. That energy goes to heat the swollen outer layers, which have expanded outward. These outer layers are large and cool. When this process happens to our star, the Sun will then be a red giant.

Not a Planet After All

After the Sun or stars like it swell to become red giants, the outer layers continue to drift out. A wind of matter flows out of the inner part of the star, blowing the outer layers outward. This wind may come in fits and starts, making various shells of matter around the star's core.

Can we see any such objects? Indeed, several have been known for hundreds of years. Already in a catalog of nonstellar objects from the 1770s put together by Charles Messier, at least the 27th object (we now call it M27), the 57th object (M57), the 76th object (M76), and the 97th object (M97) turn out to be bloated outer layers of stars. The first was discovered by Messier in 1764. In 1784, William Herschel, famous already for his discovery of the planet Uranus, thought the objects looked like tiny, hazy, greenish disks, much as the planet Uranus appeared to him. So he called them planetary nebulae.

Much later, these planetary nebulae turned out to be shells around dying sunlike stars and have nothing to do with planets. But the name has stuck. The most famous objects are these:

- M27, the Dumbbell Nebula

- M57, the Ring Nebula

- M97, the Owl Nebula

The Ring, in the constellation Lyra, is a favorite object for amateur astronomers, in part because it is high in the sky on summer nights. It looks like a tiny, hazy smoke ring to the eye peering through today's amateur telescopes. It takes an image with an electronic camera, or a long image with a film camera, to bring out the colors.

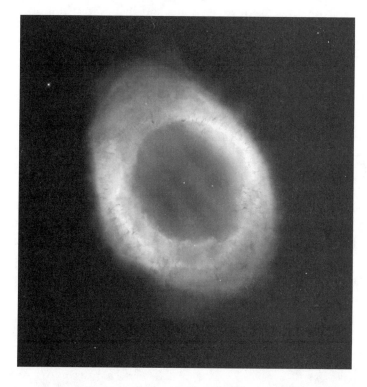

The Ring Nebula, a planetary nebula in the constellation Lyra. The Hubble Space Telescope has revealed in this image that we are really looking down a cylinder of gas. The gas was thrown off by the dying star in the center. Note that planetary nebulae do not have anything to do with planets in our solar system or in others.

(Hubble Heritage Team [AURA/STScI/NASA])

Fun Sun Facts

Planetary nebulae glow faintly greenish because of the strange conditions in them—almost a vacuum; oxygen that has lost two of its electrons contributes most of the radiation in a greenish emission line. For decades in the late nineteenth and early twentieth centuries, the green emission line was known, but its true source was not. It had been given the name nebulium, since it was found only in nebulae. The name wasn't as successful a guess as the similar name helium given to whatever caused a certain yellow line in the Sun's spectrum at eclipses. Helium, of course, turned out to be one of the basic chemical elements, while both nebulium from nebulae and coronium from the Sun's corona turned out to be from highly ionized gases.

Planetary nebulae are favorite objects for the Hubble Space Telescope's high-resolution cameras to observe because of the wonderful detail that can be seen in these relatively close objects. Hubble's images, improved over those from ground-based telescopes, have brought to prominence the various shells and other details that were smeared out previously. As a result, we now appreciate how important the episodic nature of the ejection of the gaseous shells is. And we see nonround shapes that indicate that the original stars were giving off their outer layers preferentially in certain directions.

The duration of the planetary nebula stage of a star's life is just a twinkle in the eye of the cosmos. Planetary nebulae last only about 50,000 years before they are so spread out that they are invisible. That span is only $^{50,000}/_{5,000,000,000}$, or $^{1}/_{100,000}$ (0.001 percent) of the remaining lifetime of the Sun.

We Won't Blow Up

Stars that have the mass of the Sun, or approximately its mass, see their outer layers become planetary nebulas, while their inner layers stabilize as white dwarfs. More massive stars grow hotter and denser deep inside, and the force of gravity pulling inward fuses even massive nuclei like iron into still more massive forms. That fusion of iron steals energy from the system rather than adding it, and these massive stars explode as supernovae. But the Sun doesn't have enough mass to reach that stage, so our star will never be a supernova like that.

The Crab Nebula, the remnant of a star that blew itself to shreds and whose explosion's light reached us in the year 1054 C.E.

(Hubble Heritage Team [AURA/STScI/NASA])

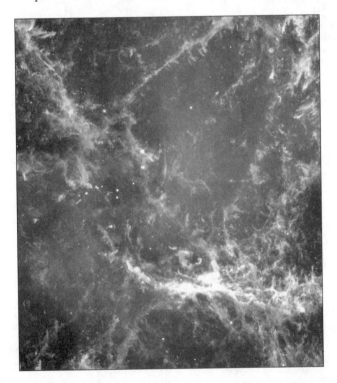

Fun Sun Facts

We do benefit from supernovae, however, even if the Sun will never be one. When they explode, they form the heaviest elements in the universe (silver, gold, and uranium, for example). The explosion spews these elements into space. When our Sun formed some five billion years ago, it formed out of a cloud of gas and dust that incorporated these elements from earlier supernova explosions. Thus each of us has within us the star stuff, to use Carl Sagan's phrase, that came from these spectacular events.

Our Ultimate Resting Place

We have already said that the outer layers of the Sun will be ejected in about five billion years and that they will fade into invisibility. That means that we will then be able to see the inner part of the Sun, the part that is left. As gravity pulls it inward, it will shrink and heat up. Eventually, the electrons will press in on each other so hard that they won't get closer together anymore. This effect is known as electron degeneracy. The pressure formed by this degeneracy counterbalances the force of gravity pulling inward. At that point, the Sun will be stable again.

Sun Words

Do not confuse the term **white dwarfs** with ordinary dwarfs like the Sun or a wide variety of stars on the main sequence of Hertzsprung's and Russell's diagram (which are yellow dwarfs or red dwarfs or, in general, dwarfs). White dwarfs have their own part of the diagram, to the lower left.

When a star stabilizes, with electron degeneracy pressure balancing gravity, it is only the size of the Earth. It has just shrunk by a factor of about 100 in diameter, which is by a factor of about 1 million in volume. The object has shrunk from being a central star of a planetary nebula, which is very hot, so it starts out very hot. That makes it blue-white hot, and the resulting objects are called *white dwarfs*.

Fun Sun Facts

In 1930, a young Indian graduate student took a steamer for London. His thinking during that long voyage changed our conception of the universe. During the voyage, Subrahmanyan Chandrasekhar (Chandra, for short) figured out that there would be a limit to the mass that a dying star could support. We now name that limit after him and realize that his work led to our conceptions of white dwarfs and other end products of stellar evolution, such as black holes.

In x-rays, as in the Chandra X-ray Observatory image shown here, Sirius B is the brighter of the pair since the hotter star, the white dwarf, radiates more x-rays. In fact, the faint light from Sirius A in this image may merely be from a leak in the filter that is supposed to pass only x-rays.

(CXC/SAO/NASA)

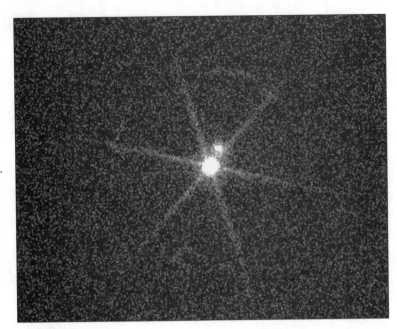

The Least You Need to Know

- ◆ In five billion years, the Sun's outer layers will swell to make it a red giant.

- ◆ The Sun will puff off a beautiful cloud of gas and dust.

- ◆ The Sun doesn't have enough mass to explode as a supernova.

- ◆ The Sun will wind up as a cold, dense cinder known as a white dwarf.

Part 3

Eclipses of the Sun

If you don't know that you are in a total solar eclipse, it is terrifying when the Sun winks out in the middle of the day. At least we now know that it is only the Moon sliding by, not anything that seriously affects life on Earth. Eclipses remain not only beautiful to experience, but also useful for discovering new things about the Sun. Recent total eclipses have occurred in Chile, India, Mongolia, Aruba, Romania, Zambia, Australia, and Antarctica, among other places. We describe these events and the ones to come. We even talk about an event that nobody now alive on Earth has seen—a passage of Venus in front of the Sun. We've had to wait over 130 years, but one will occur in 2004.

11

Who Stole the Sun?

In This Chapter

- ◆ How the Moon covers the Sun on a prescribed schedule
- ◆ The diamond ring and other beauties
- ◆ Why the corona looks round or squashed at different times
- ◆ Why you can't see the corona at lesser eclipses

For thousands of years, people have been in awe of the times when the Sun grows thin and all but vanishes in the middle of the day. Mark Twain's *Connecticut Yankee in King Arthur's Court* used the main character's advanced knowledge of such an event to gain the attention and loyalty of a less advanced civilization. (We will say more about him in Chapter 13.) The phenomena that occur at a total eclipse are so beautiful that today's eco-tourists flock in advance to places where the Sun will totally disappear. Disappointingly, though, local officials and journalists may still spread fear, acting as uninformed as the people of 2,000 years ago.

Is It a Dragon Eating the Sun?

For a random place on Earth, you would have to stand around for over 300 years before you found yourself under a total solar eclipse. But solar eclipses aren't actually very rare: Total eclipses occur about every 18 months some-where in the world, and a half dozen partial eclipses may also appear in a single year. With modern transportation, you can go to see them all.

The eclipsed Sun in the sky during the total solar eclipse of December 4, 2002, viewed over the ocean at Ceduna, Australia.

(Williams College Expedition)

But what is happening up in the sky when an eclipse occurs? From the ground, it looks as though something is taking a bite out of the Sun. For an hour or so, the Sun gets more and more covered. We now know that the "missing" part of the everyday Sun is actually the silhouette of the Moon.

Sun Safety _____

Though the Moon gradually covers the everyday Sun for an hour or more before a total eclipse, whenever any of the everyday Sun is visible, it is not safe to look at it directly. So eclipses that are only partial at best, or the partial phases of eclipses that turn out to be total, can be looked at only through special filters or viewed indirectly. We will say more later about safe ways of observing the Sun.

The Sun is about 400 times bigger than the Moon, but it is also about 400 times farther away. As a result, it takes up (scientists say "subtends") about the same angle in the sky as the Moon. Both take up about half a degree, give or take 10 percent.

The Solar Scoop _____

Your thumb, viewed at the end of your outstretched arm, subtends about 2° of arc. A full circle is 360°. Your fist, viewed at the end of your outstretched arm, takes up about 10°. You can verify that last point by checking that about nine fists take you from your horizon to the zenith (the overhead point). And you can more than cover the Sun with your thumb—though glare in the sky around the Sun often makes the bright region around the Sun larger than your thumb's apparent width.

Though you can cover the Sun's apparent disk with your thumb, all you see around your thumb is sky. Depending on how pristine and pure the sky above you is, you may see deep blue right up to the edge of your thumb, though you may have to be on a high mountain above a lot of air to see that effect. That is why most solar observatories are on high mountains. From normal altitudes and in common skies, enough dust and haze are present to make the sky brighter in the general direction of the Sun than it is off to the side.

Solar Scribblings _____

I got a surprising call recently from an advertising agency in Boston. They had seen a 20-year-old picture of me holding up my thumb against the sky, right after my family and I had landed in Hawaii following a flight to see a total solar eclipse over the ocean. You often see solar astronomers in that pose, and I was actually testing the clarity of the sky. However, the advertising agency wanted to use my picture in an advertisement showing someone admiring his thumb because video games are often played with thumbs.

At a total solar eclipse, we see the Sun's corona, which we described in Chapter 1. Why can't we see it any time that the Sun is up, just by sticking out a thumb to block the photosphere, the everyday solar surface? Well, the corona is fainter than the blue sky. So, to see the corona, we need to have the Sun up in the sky—which means that it has to be daytime—but without the blue sky. That is the circumstance of a total solar eclipse.

Solar Scribblings _____

The Apollo astronauts orbiting the Moon in 1969–1972 could look out and see the solar corona rising ahead of the Sun just before the everyday solar surface peeked out from behind the Moon. More recent robotic spacecraft orbiting the Moon also photographed this effect. But astronauts in Earth orbit don't see the effect because they see the Earth's atmosphere glowing brightly at the edge of the Earth's surface.

The solar corona rising ahead of the Sun, as seen from the Clementine *space- craft around the Moon.*

(NASA)

A total solar eclipse, when the Moon entirely blocks the solar photosphere, may last only a fraction of a second or may last as long as 7½ minutes. Only during that period is the solar photosphere entirely blocked. Therefore, only then does the blue sky vanish and the corona, normally hidden behind it, becomes visible.

The partial phases, when the Moon only partly covers the Sun, range from 0 percent to 100 percent coverage. They can last over an hour before the total phase and over an hour afterward. The Moon can be as much as about 10 percent bigger than the Sun, but once it is covering 100 percent, the eclipse is total. The extra 10 percent goes toward making the eclipse last longer.

Solar Scribblings

What is the tried and true method of getting the Sun back once it begins to disappear in an eclipse? Banging on pots and pans and on drums has been a method used for thousands of years. And it works! Never has the Sun failed to reappear. Of course, it would reappear anyway, but it's fun to make noise.

What should you do when there is a partial solar eclipse in your locality? The easy answer is to look at it, but to look safely by using a projection method or by using a special safe solar filter. Actually, though, partial solar eclipses are pretty boring. Nothing changes rapidly, and even with the sun 75 percent covered, you might not even notice the eclipse if you hadn't read about it in the newspaper (or in this book).

You should find out if totality will be visible somewhere, and do everything you can to get in the band of totality. Only there will you see the beautiful eclipse phenomena.

High Drama in the Sky

The most dramatic aspects of a total solar eclipse are the overall changes around you. When the Sun is half or more covered, you might begin to notice that the shadows on the ground become strangely sharp. That phenomenon occurs because the shadows are being projected from only a crescent-shape Sun, which is narrower in one direction, so the edges of the shadows aren't as fuzzy. As the last few percent of the everyday solar surface is covered, it quickly gets noticeably darker.

Sun Safety

Even when only 1 percent or so of the Sun is visible, what remains is still too bright to look directly at it safely. Its surface brightness is still that of the everyday Sun, even though its total brightness is reduced. So you must still use your special filters even when 99 percent of the Sun is covered.

Fun Sun Facts

During a total eclipse, the light from the Sun is only one millionth of the strength of the light from the everyday Sun. So even when 99 percent of the everyday Sun is covered, it is still 10,000 times brighter than it is during an eclipse. When 99 percent of the Sun is covered, you are only 0.01 percent of the way toward a total eclipse.

The crescent, partially eclipsed Sun before totality at the eclipse of December 4, 2002.

(Williams College Expedition)

The crescent of the remaining solar photosphere doesn't disappear all at once. First, it seems to break up into a set of bright dots known as Baily's beads. (Baily was a seventeenth-century British astronomer who described the phenomenon.) After all, the Moon's edge isn't perfectly round. Mountains stick up on the Moon, and valleys can be aligned so that a few bits of sunlight shine through them for several seconds. The photospheric light shining through those valleys makes Baily's beads.

The last Baily's bead shines so brightly compared with anything else that is visible that it is called the diamond-ring effect. The "band" of the engagement ring is often the innermost solar corona that has just begun to be visible.

The diamond-ring effect marking the beginning of totality.

Solar Scribblings

I know of at least one bride who received her proposal and engagement ring under the diamond ring in the sky. My wedding present to the couple was a photograph of the diamond ring effect that I took at the same eclipse.

During the diamond-ring's appearance, you might also notice a reddish edge to the Sun to the sides of the diamond. The reddish gas is the solar chromosphere, the intermediate layer between the photosphere and corona. It is about 1,000 times fainter than the photosphere but is still 1,000 times brighter than the corona. *Chromosphere*, of course, simply means "color sphere." As you learned in Chapter 1, it glows reddish largely because hydrogen gas emits especially strongly in that color, and you are seeing hydrogen gas silhouetted against the dark background sky.

As the diamond-ring effect fades, you see the chromosphere better for a few seconds. Then, as it is covered by the advancing Moon, the corona becomes the brightest thing in the sky. Still, it is a million times fainter than the everyday Sun. It is only about the same total brightness as an ordinary full Moon and is equally safe to look at. For however long the eclipse lasts—whether it is a second, as in 2005, or about 7 minutes, as in 2009—you will be able to look directly at the corona and see it in all its glory.

A few reddish prominences, bits of gas that are the same temperature as the chromosphere, are usually visible at the edge of the Sun. There are usually more of them at sunspot maximum than at minimum, but whether you will be lucky enough to have a bright prominence or two at the Sun's edge cannot be predicted long in advance.

What do you see during totality? Up in the sky is a dark circle surrounded by a bright halo. The circle is the silhouetted Moon, and the halo is the solar corona. A detailed look at the corona shows its spiky nature. The spikes are called coronal streamers, which show the shape of the Sun's magnetic field.

Solar Scribblings

After every eclipse, you hear stories of someone who forgot to remove the camera's special solar filter during totality. If you try to photograph through or look through the solar filter, you won't see anything at all! The solar corona is too faint to be seen through the filters that are needed for partial phases.

During totality, it is quite dark around the Sun, since the blue sky has gone away. You are in the cone of the shadow of the Moon as it falls upon the Earth. But if you look at the horizon in any direction, you are looking outside that shadow cone, and you can see regions where the Sun is shining. Because you are looking through lots of air, only the reddish light comes through to you. Thus, you see a 360° sunset effect all around you. Being outside during this stage is exciting and dramatic. All the beauty and drama is lost when you merely view an eclipse on television or on the Internet.

Squashed and Flat

If you were a heart surgeon and were told that you could see inside a human heart, but only if you traveled halfway around the world, and then that you would get only a minute or so to observe it, you surely would go. You would gather lots of colleagues and cameras and do what you could during that minute or so. The case for observing eclipses is similar. You may have to travel to get into the path of totality, but the observations that you can make and the data that you can collect during the brief interval of totality can be studied later. And if a year or two later, you had another opportunity, you'd go again, wouldn't you? You wouldn't say, "Well, I looked inside a heart last year." You'd be glad for the extra time. The same is true with eclipses. Observing one for a brief interval still leaves lots to do at the next eclipse.

Another reason for observing every possible eclipse is that the Sun changes from day to day and from year to year. The sunspots on the solar photosphere show where tubes of magnetic field leave and re-enter the Sun. They show where the Sun's magnetic field is strongest. The magnetic field extends outward into the corona and holds the coronal gas in place.

The shape of the corona changes with the sunspot cycle. Here are several eclipses photographed with special cameras of the High Altitude Observatory, Colorado, that minimize the bright inner corona to make the coronal shape more apparent.

(High Altitude Observatory/ National Center for Atmospheric Research)

Solar Scribblings

Researchers have been trying for decades to make fusion energy in power plants on Earth. Among the biggest obstacles is the fact that no mechanical wall can withstand the high temperatures. Researchers' best tries use magnetic fields to hold the hot gas in place. In the Sun, magnetic fields hold the very hot coronal gas in place. The lessons we learn from the Sun's constraining those hot gases may lead to an improved understanding of the physical laws that can help us master fusion for peaceful purposes on Earth.

The sunspots, and the magnetic fields all over the Sun, go through an 11-year cycle. The shape of the solar corona, therefore, also goes through an 11-year cycle. During low points of the sunspot cycle, only a few coronal streamers are visible, and they stay mainly near the Sun's equator. But during high points of the sunspot cycle, there are more coronal streamers at high solar latitudes. When projected against the sky, they seem to stick out in all directions. The overall shape of the corona seen at eclipses then becomes round. So, over an 11-year cycle, the shape of the corona goes from apparently squashed at sunspot minimum, to round at sunspot maximum, and back to squashed again. We had round coronas for the eclipses from 1999 to 2002, and now we will have minimum-type, squashed coronas through perhaps 2007.

The End of Glory

After the seconds or minutes of totality, you may see a brightening and a bit of reddish chromosphere on the solar limb. Then the diamond-ring effect appears again. All too quickly, the sky brightens and the total phase of the eclipse is over.

The partial phases at the end last about as long as the partial phases did leading up to totality. But they are much less well observed and much less photographed. After the glory of totality, the partial phases are anticlimactic.

The Least You Need to Know

- A total solar eclipse occurs somewhere in the world about every 18 months.

- The exciting times of a solar eclipse are Baily's beads, the diamond-ring effect, the chromosphere, and the corona sequences.

- The corona looks round at the maximum of the sunspot cycle and elliptical at sunspot minimum.

- Only at a total solar eclipse does it get dark enough to see the corona.

12

Saros and Cycles

In This Chapter

- ◆ The Moon's shadow determines the type of eclipse

- ◆ Total eclipses occur about every 18 months

- ◆ Annular eclipses occur about as often as total eclipses

- ◆ Months and years mesh in an eclipse cycle

- ◆ Eclipses move around the world

Though the eclipse in Australia in 2002 lasted only 32 seconds, some people enjoyed almost 7 minutes of totality in Baja California on July 11, 1991. How can they know when the next very long eclipse will occur? They just add 18 years 11⅓ days, giving July 22, 2009. The third of a day allows the world to spin from Mexico to China. The clockwork of cycles of the Sun and the Moon give such unusual regularities.

Almost Too Perfect a Fit

The Moon orbits the Earth every 27⅓ days, if you could look down on it from high above the solar system. But while it does so, the Earth moves ahead in its orbit. So it takes the Moon 29½ days to catch up to the same position on Earth with respect to the Sun. Since the phases of the Moon depend on the angle that sunlight hits it, lunar phases repeat with this 29½-day period.

The Moon's orbit around the Earth isn't perfectly round. Sometimes the Moon is about 10 percent closer to Earth and sometimes 10 percent farther away from its average distance. So sometimes the cone of the Moon's shadow hits the Earth, while at other times it doesn't quite reach.

The total eclipse, top, is viewed with the naked, unfiltered eye or camera. The annular eclipse, bottom, must be viewed only through a special filter that blocks all but about ¹/₁₀₀,₀₀₀th of the sunlight.

The cone of the Moon's shadow reaches the Earth in a total eclipse but doesn't reach it in an annular eclipse.

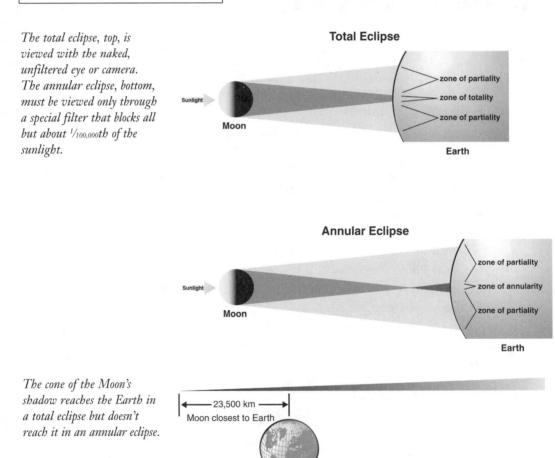

Total and Annular Solar Eclipses

When the cone of the Moon's shadow hits the Earth, we have a total solar eclipse. People standing within that cone and looking back toward the Moon see it blocking the everyday Sun entirely. The sky turns black, and we see the solar corona.

When the cone of the Moon's shadow doesn't quite reach Earth, we have an annular solar eclipse. People standing within the extension of that cone and looking back toward the Moon see a ring of bright sunlight around it. *Annulus* is a word (from the Latin) for "ring," so the event is called an *annular eclipse*. Even when only a tiny rim of sunlight remains visible, even down to a fraction of a percent, enough sunlight remains to keep us from seeing the corona.

Sometimes the point of the cone of the Moon's shadow just barely reaches the Earth at the part of the Earth's surface that is directly facing the Sun. On these rare occasions, we have an *annular-total eclipse*. In 2005, an annular eclipse will cross the Pacific Ocean. For only a few miles in the middle of the path, the Moon's shadow will reach Earth and the eclipse will be total. We hope that cruise ships will take hundreds of passengers out to that point, though certainly many fewer people will see totality than the tens of millions of people in Mexico City and elsewhere who saw totality at the 1991 eclipse.

Since both total eclipses and annular eclipses occur only when the Moon is centered on the Sun as seen from Earth, they are called *central eclipses*.

 Solar Scribblings

In a typical century, there are 70 total eclipses, 72 annular eclipses, and 6 annular-total eclipses. The longest solar eclipse was over 7 minutes, and the longest annular eclipse was over 12 minutes.

CAUTION **Sun Safety**

Except during totality of a total solar eclipse, you need a special solar filter if you want to look at the Sun. Never look unprotected at a partial or an annular eclipse.

 Sun Words

An eclipse of the Sun in which an annulus (a ring) of bright everyday sunlight remains visible around the Moon is an **annular eclipse**.

An eclipse of the Sun that is annular for the ends of its path but total in the middle is an **annular-total eclipse**. A total or annular solar eclipse is a **central eclipse**.

But the Sun and Moon don't have to be perfectly aligned for the Moon to partly block the Sun for people on Earth. If the point of the cone of the Moon's shadow passes above the Earth's North Pole or below the South Pole, it never hits the Earth. Then we on Earth see no more than a partial eclipse. This circumstance happens more often than do total or annular eclipses. We can have as many as five solar eclipses each year, but most of them are only partial, and the beautiful phenomena of a total eclipse don't appear. Some years have as few as two solar eclipses.

Looking up through a special solar filter at the partial phase of an eclipse.

(JMP; Williams College Expedition)

Eclipse Seasons

The Moon's orbit is tilted by 5° compared with the plane of the Earth's orbit around the Sun. So most months, the Moon merely passes above or below the Sun in the sky at the time of new moon. The Moon's orbit passes through the plane of the Earth's orbit around the Sun at two points on opposite sides of the orbit. Those points are called *nodes*. Only when the Moon and Earth are both near one of the two nodes do we have an eclipse. These times repeat every 173 days, once on each side of the orbit. These times are *eclipse seasons*. Twice that time makes 346 days, which we call an *eclipse year*.

Sun Words _____

A **node** is a place where two curves cross or a wave doesn't change over time.

An **eclipse season** is the time of year when the Earth and the Moon are close enough to the nodes to potentially have an eclipse.

An **eclipse year**, 346.62 days, is the period with which the Earth passes through opposite nodes.

The time the Moon takes to go from one of its nodes through the other and back again is called the *nodical month*. Since eclipses happen according to this schedule, the intervals are also called *draconic months*. *Draconic* refers to the mythical Chinese dragon that ate the Sun to make eclipses.

The central part of the shadow cone is called the *umbra*. A lighter shading in the diagram shows locations where the Sun is only partially shadowed by the Moon. From those locations, part of the Sun is visible; people there have a partial eclipse. These locations are the shadow's *penumbra*. In an annular eclipse, Earth passes beyond the point of the umbra into a region sometimes known as the *antumbra* (for "anti-umbra").

Sun Words _____

Nodical months, or draconic months, 27.21 days, are the intervals between the Moon's return to the same node.

The **umbra** is the completely dark part of a shadow.

The **penumbra** is the set of places that are only partially shadowed.

The **antumbra** is the continuation of the umbral cone beyond its point.

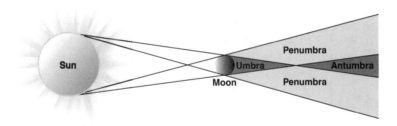

The parts of the Moon's shadow: umbra, penumbra, and antumbra.

Over and Over, but Not Quite the Same

You have learned that the Moon misses blocking the Sun most months because its orbit is tilted. But several times a year, the Earth is close enough to one of the nodes when the Moon passes through it. Then we have a total or annular solar eclipse.

Whenever three astronomical bodies are in a line, we have a *syzygy*. (I love that "yzy," which stumps many people when they are playing the word game Ghost.) Since the Earth goes around the Sun in 365¼ days and the Moon goes around Earth in 27⅓ days, there is no particular reason for the three bodies to come back in a line at regular intervals. But through a fluke, a syzygy of the Sun, the Moon, and the Earth occurs when the

Sun Words _____

A **syzygy** is a lineup of three astronomical bodies.

Moon has gone around the Earth a certain integral number of times—that is, an integer like 1, 2, 3, rather than a fraction or decimal—and the Earth has gone around the Sun a different integral number of times. It is just a lucky coincidence. That length of time works out to 18 years 11⅓ days, though it can be 18 years plus 10⅓ days or 12⅓ days, depending on how the leap years fall in the interval.

Fun Sun Facts

Partial eclipses can be fun to see but are usually too bright to view without special glasses. But sometimes at sunrise or sunset, haze on the horizon makes the sun sufficiently dim so that it can be looked at or photographed. When this happens for a partial eclipse, it is nice to see the horns of the Sun as it goes down. In Australia in 2002, some people went so close to the end of the eclipse path that the totally eclipsed Sun was only a Sun's diameter above the horizon. I have never seen a photograph of a fully eclipsed Sun in the process of rising or setting—nor would I ever take the chance of taking such a photograph myself, given a lower percentage of interfering clouds where the Sun is higher in the sky.

Sun Words

The **saros** is the interval of 18 years 11⅓ days (plus or minus a day on the calendar, depending on leap years) over which eclipses repeat.

This interval is known as the *saros*, a name chosen by Edmond Halley (of comet fame), who mistakenly thought that the ancient Babylonians had used the name for such an interval. (The Babylonians had actually used the term for something else that was 18½ years in duration.) So when there is an eclipse, one saros later there is also an eclipse of about the same duration. And that eclipse is one third of the way around the globe from where the earlier eclipse was.

Solar Scribblings

223 lunar (phases) months (@29.5306 days) = 6,585.32 days

242 nodical (draconic) months (@27.2122 days) = 6,585.36 days

So the moon is "new" while it is again going through one of its nodes.

19 eclipse years = 6585.78 days

239 lunar-orbital-shape months (@27.5546 days) = 6585.54 days

In addition to the Sun and the Moon going back to the same place at the same time, even the alignment of the Moon's elliptical orbit goes back to nearly the same place. This alignment follows the *anomalistic month*. So if the Moon is relatively close to Earth, it will be just as close after exactly one saros interval has passed. Thus, if you have a long eclipse, one saros later there will be another long eclipse. And if you have a short eclipse, unfortunately, it will still be short a saros later. (However, the short ones might be even more exciting than the longer ones. After all, you might have only a second or so after the beautiful diamond-ring effect at the beginning before another diamond-ring effect appears.)

Sun Words

An **anomalistic month**, 27.55 days, is the period between perigees—the closest approaches of the Moon to the Earth.

Within a saros interval of 18 years 11⅓ days is a series of total eclipses. For example, between the long eclipse of 1991 and the long eclipse of 2009, eclipses occurred or will occur in the following years:

- ◆ **1992**—A short eclipse over the south Atlantic ocean

- ◆ **1995**—India South Asia, with a peak of 2 minutes 10 seconds; observed by many for 30 seconds in India

- ◆ **1997**—Mongolia and Siberia, with a peak of 2 minutes 50 seconds

- ◆ **1999**—Europe, peaking over Romania, with a peak of 2 minutes 23 seconds

- ◆ **2001**—Southern Africa, up to 4 minutes 57 seconds over the ocean, 4 minutes 35 seconds at landfall, and 3 minutes 38 seconds in Lusaka, Zambia

- ◆ **2002**—Southern Africa and Australia, peaking at 2 minutes 4 seconds over the ocean between them

- ◆ **2003**—Antarctica, peaking at 1 minute 57 seconds

- ◆ **2005**—A mostly annular eclipse, turning total briefly over the Pacific Ocean

- ◆ **2006**—Africa and Turkey, peaking at 4 minutes 7 seconds

- ◆ **2008**—Greenland, Siberia, Mongolia, and China, peaking at 4 minutes 27 seconds

- ◆ **2009**—India to China, peaking at 6 minutes 39 seconds

The following 18 years will have a series of eclipses of about the same lengths in the same order.

If we follow a pattern of eclipses from saros to saros, you see that eclipse paths move north or south across Earth, starting near one pole, gradually moving through the temperate latitudes and equator, and moving off the other pole some 1,200 to 1,500 years later.

Eclipses and World Travelers

I saw my first solar eclipse within 10 miles of where I was living at the time. So I was enthralled by the eclipse itself, not the tourist aspects. But since that time, the 35 solar eclipses I have seen have taken me all over the world.

Basically, the Moon's shadow is a tapering cone. If you cut a cone with a plane perpendicular to the axis of the cone, you get a circle. But if the plane is tipped, you get an ellipse. The cone of the Moon's shadow is being intercepted by the curved surface of the Earth, so the actual shape of the shadow on the Earth is complicated but is close to an ellipse.

As the Earth and the Moon move in space around the Sun, and as the Earth rotates, the elliptical intercept of the shadow and the Earth sweeps across Earth from sunrise to sunset. So the path is only up to a couple of hundred miles across, but over 10,000 miles long. The paths of eclipses in the forthcoming period are shown in the following figure.

Total eclipses between 2001 and 2025.

(F. Espenak, NASA's Goddard Space Flight Center)

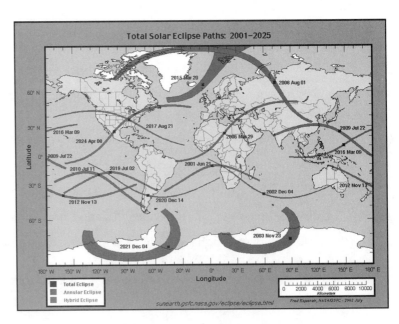

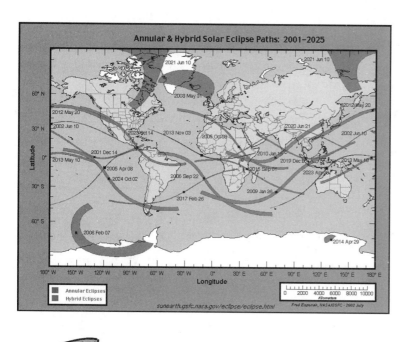

Annular eclipses between 2001 and 2025.

(F. Espenak, NASA's Goddard Space Flight Center)

Sun Safety

Here's how you can look at the Sun:

Every day: You need a special filter (the same special filter as during the partial phases).

Partial phases: You need a special filter.

Totality: Look directly, without a filter.

Diamond ring effect: Put your filter back on.

Final partial phases: You need a special filter.

The cone of the Moon's shadow actually moves in space at about 2,000 miles per hour. If the Earth weren't *rotating*, the shadow would pass over the Earth at that speed. But the Earth is rotating—that is, spinning on its axis, as opposed to *revolving* around the Sun.

Note that if you are standing on the North Pole, the Earth may be spinning but you aren't moving. So, at or close to the poles, an eclipse moves at the full speed of the shadow in space. But if you are at or near the equator, Earth is turning at a speed of its circumference divided by 1 day, or 25,000 miles divided by 24 hours, which is roughly 1,000 miles per hour. Since

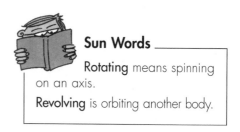

Sun Words

Rotating means spinning on an axis.

Revolving is orbiting another body.

the Earth's surface is rotating in the same direction that the eclipse ellipse is moving, people at the equator are partially keeping up with the eclipse's motion. The speed of an eclipse that occurs at the equator at noon drops down to 1,000 miles per hour. A supersonic plane can keep up with that speed, at least for a while. But ordinary passenger planes—though they may be useful for getting above the clouds—don't go fast enough to keep up with an eclipse.

> ### Fun Sun Facts
>
> The Moon's shadow can be as much as 269 km wide. Because of the Earth's curvature, it can cover a width of over twice that on Earth's surface.

The maps show not only the locations of the centers of the eclipse paths, but also their widths. Because of the curvature of Earth's surface, the paths widen near the poles; some widening is also artificially apparent because of the vagaries of the Mercator projection used.

The durations of totality stay pretty close to the maximum even when you are quite a bit off the center line. Only near the very edges of the path do the durations go down substantially.

Solar Scribblings

If you stand in one spot, a total eclipse appears overhead about every 300 years. But the statistics are more favorable in some spots. One location in Angola saw eclipses in both 2001 and 2002, with an interval of less than 18 months. Even slight traveling lessens the interval. Statistics from a book cataloging all eclipses in England in 3,000 years from 1 C.E. to 3000 C.E. showed that in a 50-mile area around London, the average gap between total eclipses is 330 years, while in a similar area around York, the average gap is only 156 years.

The Least You Need to Know

- When Earth is close enough to the Moon to intercept its shadow, we have a total eclipse.

- Total eclipses and annular eclipses each occur about every 18 months.

- Eclipses of the same duration recur at the saros interval of 18 years 11⅓ days.

- Eclipse paths are no more than a few hundred miles wide, even though they are over 10,000 miles long.

- You must watch partial or annular phases of eclipses through special filters, but no filter is needed for totality.

- Total eclipses cross any bit of Earth every 300 years, on average, but traveling makes totality more frequent.

13

Helium: Only on the Sun

In This Chapter

- ◆ Eclipses have been significant for thousands of years

- ◆ Helium, since found on Earth, was first detected in the Sun

- ◆ Coronium turned out to be highly ionized atoms, not a new element

- ◆ The magnetic field is important for shaping the corona

- ◆ The Sun is an excellent laboratory

The Sun tootles away year after year and century after century, converting hydrogen to helium and pumping out the energy to warm us on Earth. How can we re-create on Earth the conditions that we find in the Sun, to provide limitless power? What other lessons can we learn by studying the Sun, a celestial laboratory placed close enough for us to study in detail, but far enough away that it rarely harms us?

Historical Eclipses in Science, Literature, and Politics

Some of the most widely mocked astronomers were Hsi and Ho in ancient China, about 2000 B.C.E. They were said to have been so drunk that they failed to predict an eclipse. The emperor, supposedly, had their heads. The real story, though, is probably that Hsi-Ho was a name for a sun deity whose job included preventing eclipses. Did drunk astronomers really fail to predict the eclipse in 2134 B.C.E.? Probably not.

October 16, 1876 B.C.E.

Some of the oldest astronomical records are found in China on tortoise-shell chips. These "oracle bones" or "dragon's bones" sometimes have statements that may refer to eclipses. A Jet Propulsion Laboratory scientist and a UCLA professor of Chinese have used some of these oracle bones to date one eclipse to 1302 B.C.E.

Historical eclipses from long ago are particularly important because the Earth does not rotate at an absolutely constant rate. Sometimes it goes a tiny bit faster and sometimes a tiny bit slower, a result in part of how concentrated the Earth's mass is at various distances from its center. Changes in Antarctic ice, for example, can affect the speed of the Earth's rotation.

Eclipses are a sensitive way of measuring the speed of the Earth's rotation. We can use the Earth's current speed of rotation to predict where eclipse paths were hundreds or thousands of years ago. A total eclipse is such a spectacular phenomenon that clues to whether it occurred can sometimes be found in ancient records of various sorts. If the Earth's rotational speed has varied faster or slower than average, the observed eclipse path will have shifted east or west from the predicted path.

F. Richard Stephenson and his colleagues at the University of Durham in the United Kingdom have been leaders in interpreting ancient eclipses. They have found a slowing of the Earth's rotation over millennia. Since one Earth rotation is one day, they are basically measuring the length of the day. If the record of the 1876 B.C.E. eclipse is correct—which it might not be since it was found in a book from the sixth century B.C.E., over 1,000 years later—it would indicate that a terrestrial day was 0.07 second shorter in 1876 B.C.E. than today.

June 5, 1302 B.C.E.

How would you interpret "Three flames at the sun, and big stars were seen"? A team studying historic eclipses found this statement on "dragon's bones." The description could match the total eclipse of June 5, 1302 B.C.E., with the "flames" corresponding to coronal streamers.

585 B.C.E.

Herodotus, a Greek whose history has greatly shaped our views of ancient times, described in 430 B.C.E. how the total eclipse of May 28, 585 B.C.E., was visible during a war between the Lydians and the Medes. Herodotus wrote that the Lydians and the Medes ended the war as a result of the eclipse. What is more questionable in this report about 150 years after the fact was the statement that the Greek scientist Thales of Miletus had predicted the eclipse. From his careful study of ancient eclipses, Stephenson thinks that this time scale was too early and that, from what we know about the evolution of scientific abilities, there is no way that Thales could have predicted an eclipse that early. In our current evaluation, therefore, the story seems to be myth.

Lunar Eclipse of February 29, 1504

Christopher Columbus used this eclipse when he was stranded in Jamaica. Though it was merely a lunar eclipse, the story was picked up and associated later with solar eclipses, so we include it here. Columbus and his crew were suffering for lack of supplies. Fortunately, Columbus had a table of astronomical events and positions with him. It showed that there would be a lunar eclipse on February 29, 1504. He arranged to meet with the local chiefs during the eclipse and said he would take the Moon away. When they were sufficiently awed and scared, they arranged for food to again be brought to his crew.

April 22, 1715

Edmond Halley predicted the path of this eclipse, which passed over England from southwest to northeast. Ten years earlier, he had analyzed the paths of several comets in the sky and deduced that some of them represented the same comet returning. We still call that comet Halley's comet. Halley had sponsored Newton in producing his major work and in deriving the law of gravity. Halley continued to use the newfangled law of gravity for his predictions.

The map drawn for Halley of the 1715 eclipse was the first of its type, though maps are now routinely drawn to show the paths of eclipses over the Earth's surface. Space satellites now even peer down, showing the umbra moving across the Earth's surface.

After the 1715 eclipse occurred, a corrected map was issued, including the predicted path for the 1724 eclipse that would also pass over England.

Along with his map of the actual path of the 1715 eclipse, Halley added predictions for the path of the 1724 eclipse, shown here.

(Jay M. Pasachoff)

October 27, 1780

An eclipse was predicted for North America, but the problem for Harvard astronomer Samuel Williams was that the prediction was across enemy lines. The enemy, of course, was the British. Nonetheless, Williams set out for Penobscot Bay, in what is now Maine but was then Massachusetts, with a group of students and colleagues.

The delay of getting across the British lines turned out to be part of the problem, given also that there were errors on the maps they used. They wound up slightly outside totality. They did see the phenomenon that later was seen by Baily and named after him (Baily's beads).

May 15, 1836

English astronomer Francis Baily noted the beads of sunlight that appeared on the edge of the Moon during an annular eclipse that he observed from Scotland. He described them so well and so enthusiastically that we still talk at each eclipse of Baily's beads (though his name is misspelled as often as it is spelled correctly, it seems).

July 16, 1806

Tenskwatawa, a Shawnee prophet who was a leader of Native Americans in Ohio and Indiana about 200 years ago, acted as Columbus had. He threatened William Henry Harrison, who was then governor of the Indiana Territory and who later became president of the United States, to "blacken the face of the sun." When a large crowd watched Tenskwatawa at the eclipse, Harrison lost his attempt to diminish Tenskwa-tawa's influence.

1879 with No Eclipse, and June 21, 528

In Mark Twain's nineteenth-century story, the hero of *A Connecticut Yankee in King Arthur's Court* awoke in ancient England and fortunately knew …

> that the only eclipse of the sun in the first half of the sixth century occurred on the 21st of June, A.D. 528 … and began at 3 minutes after 12-noon. I also knew that no total eclipse of the sun was due in what to *me* was the present year—i.e., 1879.

He also remembered …

> how Columbus, or Cortez, or one of those people, played an eclipse as a saving trump …, and I saw my chance …. "Go back and tell the king," our hero said, "that at that hour I will smother the whole world in the dead blackness of mid-night; I will blot out the sun, and he shall never shine again; the fruits of the earth shall rot for lack of light and warmth, and the peoples of the earth shall famish and die, to the last man!"

Our narrator became a hero when …

> It got to be pitch-dark, at last, and the multitude groaned with horror to feel the cold uncanny night breezes fan through the place and see the stars come out and twinkle in the sky. At last the eclipse was total, and I was very glad of it …. I said, with the most awful solemnity: "Let the enchantment dissolve and pass harmlessly away!" When the silver rim of the sun pushed itself out, a moment or two later, the assemblage broke loose with a vast shout ….

1885

Another famous old story was *King Solomon's Mines*, by H. R. Haggard, from 1885. In it, the hero gains influence by threatening to "darken the sun to-morrow." But Haggard allowed an hour for totality, so he obviously had never seen one himself.

Only on the Sun

When Louis J. M. Daguerre invented photography in 1839, imaging the Moon was one of his prime goals. But it took until 1851 for the first astronomical photographs to be made successfully. The first photographs of the Sun were taken about then, including even an image showing the corona at the 1851 eclipse.

British amateur astronomer Warren De la Rue and Italian cleric Father Angelo Secchi took photographs of the 1860 eclipse from different locations. At that time, it was still debated whether the corona and other phenomena seen were on the Sun or in the Earth's atmosphere. From the fact that the prominences looked identical from different locations, scientists concluded that they were really on the Sun, since they were looking through different regions of the Earth's atmosphere.

> **Fun Sun Facts**
>
> Father Secchi wrote a wonderful book about the Sun. A NASA solar satellite to be launched into space will be named after him: Sun Earth Connection Coronal and Heliospheric Investigation (SECCHI). It is to be a component of the Solar Terrestrial Relations Observatory (STEREO).

Scientific study of the sun was greatly enhanced in the nineteenth century by the discovery of spectroscopy. Newton had shown in the seventeenth century that white light from the Sun could be broken down into its component colors and that the colors could be reassembled to make white light again. But it took until the inventions and careful work of the German optician Joseph Fraunhofer in the early nineteenth century to transform spectroscopy into a useful tool for solar research.

An eclipse drawing of the eclipse of 1860, from a book by Father Secchi.

(Jay M. Pasachoff)

In Chapter 5, you learned that Fraunhofer spread out the spectrum of the solar photosphere in 1814. You learned how he labeled the strongest spectral lines with letters, especially C, D, and H.

The first time a *spectroscope* went to an eclipse was on the expedition to India in 1868 run by Pierre Jules César Janssen of Paris. Janssen could look only with his eyes at the spectrum; photography was not advanced enough to be able to record images of spectra during the brief interval of totality. The addition of a camera to record the spectrum transformed the spectroscope into a *spectrograph*, something that had already been accomplished by French scientist Léon Foucault.

Sun Words

A **spectroscope** is a device that you can look through to see the spectrum. A **spectrograph** is a device that records the spectrum on film or digital media.

Helium on the Sun

Janssen watched with amazement as the diamond-ring effect vanished and the chromosphere was briefly visible. He could see a bright yellow line in the spectrum at that time. He thought it was located at about the position of the D lines in the photospheric spectrum—a pair of lines, D_1 and D_2, that come from sodium.

Janssen realized that the yellow line was so bright that he could potentially study it even without the eclipse. Soon he opened the slit of his spectroscope and looked at the lines in a more leisurely manner. He discovered that the line, though yellow, was not exactly at the position of the D_1 and D_2 lines from sodium. He therefore called the line D_3. He said it was from "helium," since Helios is the Sun god in Greek mythology and it was an element found only on the Sun.

Coincidentally, at about the same time, Norman Lockyer in England, who hadn't been at the eclipse, decided to examine the Sun's edge with a spectroscope. He also discovered the D_3 line of helium. It took another 27 years, until 1895, for helium to be isolated on Earth. We now realize that it is element number 2 and makes up about 10 percent of all the atoms in the universe.

Coronium in the Corona?

Scientists took spectrographs to the eclipse in 1869. They realized that during totality, there was a bright green emission line in the spectrum. That showed them that hot gas was present in the corona, itself a major discovery. But they didn't know what the gas was. It was soon called coronium, since it was apparently (for the people of that time) found only in the corona. Charles Young, a Princeton professor and leading American expert on the Sun, wrote in his 1895 book on that topic, "As to the substance … we have no knowledge as yet, though the name 'coronium' has been provisionally assigned to it, and the recent probable identification of 'helium' in terrestrial minerals gives strong hope that before very long we may find coronium also."

Over the following years, other coronal emission lines were found. The second strongest in the visible occurs in the red and is known as the coronal red line, to go with the coronal green line.

The coronium problem proved much harder to crack than the helium problem. The key came from studies in the 1930s of novae, explosions on star surfaces. The spectra of two novae showed emission lines that the brilliant Swedish spectroscopist Bengt Edlén identified with unusual transitions in very hot iron atoms. Edlén went on in 1939 and the following years to show that the several coronium lines that had been observed since 1869 came from very hot versions of several elements, especially iron, calcium, and nickel. His studies started with cooler gases and extrapolated to the hotter ones that he couldn't study in the laboratory but that he showed existed on the Sun. So, the coronium problem took 70 years to solve. There was no such thing as "coronium," after all! However, many lines of evidence support Edlén's deduction that the solar corona is very hot and that "coronium" is explained by these hot gases of ordinary elements.

> **Solar Scribblings**
>
> When Leon Golub of the Harvard-Smithsonian Center for Astrophysics sent up a rocket for the eclipse in 2001 that observed nickel-17 (nickel that has lost 16 electrons out of its quota of 28), he was observing gas at a temperature of over 5 million°F (3 million°C).

Solar Scribblings

Other indications showed that the corona was very hot even before Edlén's spectroscopic and theoretical work. Aside from the emission lines, the inner corona doesn't show any absorption lines. Yet if the coronal gas were cool, it should reflect the absorption lines in the photospheric spectrum. The solution to that conundrum is to realize that the coronal electrons are moving very fast—that is, they have high temperature. From the smearing of the absorption lines, we can deduce that the coronal gas must be millions of degrees. But the experiment is very hard to do.

In particular, the coronal green line comes from iron that has lost 13 of its 26 electrons. The coronal red line comes from iron that has lost 9 of its 26 electrons. The hotter it is, the more electrons are given enough energy to separate from the atom.

An image of Ni XVII—that is, 16-times-ionized nickel—taken from a rocket flight.

(Leon Golub, SAO)

A Laboratory in the Sky

Often objects in the sky present us with new situations that we earthlings haven't experienced. We try to use those unusual objects and events to understand more about the universe in which we live.

Magneto-hydro-dynamics

When we look at the corona at an eclipse, we see streamers extending into space. Some are called helmet streamers, in which the base is wider and the top is pointy, like an old Prussian helmet that we might see in the history books or in a movie. These streamers are held in place by the Sun's magnetic field.

When we look close to the limb of the Sun, we see loops of gas in the corona. These coronal loops are also held in place by the Sun's magnetic field. The TRACE space-craft makes particularly detailed images of them.

These magnetic formations mimic lines of magnetic force traced by iron filings held near a magnet. Some scientists have calculated the magnetic field at a theoretical sur-face arbitrarily placed near but above that of the Sun, and have shown how the mag-netic field extends into space. They have even predicted what the corona will look like during an eclipse, though the predictions have not been entirely successful.

On Earth, we are looking for ways of generating power for business and home use. We mainly now use fossil fuels, which lead to greenhouse warming (which we will discuss later) and to dependence on oil from the Middle East and other politically undesirable places around the world. Nuclear power is free of the consequences of global warming, so it would be desirable from that point of view if it could be even more widespread than the approximately 20 percent of energy it now generates in various countries. But concerns about weapons proliferation, the economics of the nuclear cycle, and long-term hazards of nuclear waste have left the nuclear industry in a moribund state.

Solar Scribblings

The shape of the magnetic field calculated for the corona sticks out like a ball with spiky hair sticking out to outline the coronal streamers. These calcula-tions are therefore known as hairy-ball models.

The Holy Grail of power has for a while been nuclear fusion, the same process that fuels the Sun and stars rather than the nuclear fission that makes today's nuclear power plants work. One of the major problems with fusion is that it works by putting together two nuclei; each nucleus has a positive charge. The nuclei thus repel each other, and they have to be forced close enough to fuse. That requires high temperatures—millions of degrees—and suffi-ciently high densities. Where do we find such condi-tions in nature? Of course, in the Sun and stars.

As we mentioned in Chapter 1, the fusion reactor at the Sun's core is 93 million miles away from us and is held in place by gravity at a safe distance from us. Other million-degree gas at the Sun is in the corona, where it is held in place by magnetic fields. For fusion to take off on Earth, we must control the fusion process and keep hot, dense gas in place. No actual container can hold such hot, dense gas. The main ideas for overcoming the obstacle involve a device called a tokamak, which uses magnetic fields to make a "magnetic bottle" to hold in the gas. Unfortunately, the magnetic bottles are not stable, and they come apart in a fraction of a second. By studying the laws of physics through observation of hot gases held in magnetic fields in the Sun, we may learn more about how to control fusion on Earth.

The Solar Scoop

Tokamak is a Russian acronym for "toroidal magnetic chamber with an axial magnetic field." That is, a tokamak is a bagel-shaped (doughnut-shaped) holder (thus, a toroidal holder) with a magnetic field centered on the toroid's axis—the line perpendicular to the plane of the bagel and at the center of its central hole. The next-generation tokamak is to be ITER, the International Thermonuclear Experimental Reactor, a $5 billion project.

Fun Sun Facts

In the Sun, a nucleus of ordinary hydrogen fuses with other hydrogen nuclei to build up to helium, which incorporates four hydrogens. At still lower temperatures, deuterium fuses to make helium. Ordinary hydrogen is just a single proton, while deuterium—also known as "heavy hydrogen"—is a proton and a neutron bound together. Types of failed stars called brown dwarfs have recently been discovered by the dozen. They don't have enough mass to get hot enough inside to fuse ordinary hydrogen, but they do fuse deuterium.

Gas that is ionized—that is, separated into its positive particles (nuclei) and negative particles (electrons)—is known as plasma, as we described in Chapter 6. The Sun and stars are made of plasma. The study of things moving is called dynamics, and the study of the motion of fluids (and other things that flow, including gases) has long been called hydro- from the study of moving water. When you add the term magneto- for the magnetic field, you get magneto-hydro-dynamics, or *magnetohydrodynamics*, often simply called *MHD*. The laws of MHD govern the hot, ionized gas in the Sun.

Sun Words

Magnetohydrodynamics, or **MHD**, is the study of the motion of ionized gases (plasmas) in magnetic fields.

Forbidden Lines

Spectral lines—whether emission lines or absorption lines—come from electrons changing orbits in atoms. For each orbit, we say that each electron is on a certain "energy level" and that the energy levels are spaced out in certain ways rather than being continuous. In 1913, Niels Bohr worked out the basic idea: Each electron stays on a certain energy level and can jump to another energy level, but it cannot take intermediate values or energy. The situation is similar to climbing stairs: You can be on one step or another, but you cannot hover between them. Whenever an electron changes from one energy level to another, the difference in energy corresponds to a spectral line of a corresponding color.

When the laws of quantum mechanics were worked out in detail in the late 1920s, some rules turned up that governed which pairs of energy levels could give off and take up an electron in a given jump. These rules showed a certain set of jumps that are *permitted*. These permitted jumps corresponded to *permitted lines*. All other transitions were called *forbidden*.

Sun Words

Permitted lines are spectral lines that are common (we say "allowed") from transitions between two energy levels of atoms. **Forbidden lines** are spectral lines that would occur so rarely that other circumstances in atoms, like collisions, would keep them from occurring. Thus, they appear only in gas of extremely low density.

Fun Sun Facts

The solar corona has only about a billion particles in each cubic centimeter of gas. The Earth's atmosphere has trillions of times more particles in each cubic centimeter. The density of the solar corona is lower than we can make in laboratories on Earth.

But forbidden lines are not absolutely impossible; they are merely of low probability. If an atom is sitting on one energy level, it spontaneously can drop to a lower energy level, just as a ball rolling off a top step can fall to a bottom one. But if the ball has a low probability of rolling off the step, someone might kick it off first, and it wouldn't necessarily go to the lower step it would have fallen to. (It could go to a higher level or even out of the atom entirely.) The same thing happens on the Sun and on stars. On the Sun as well as on the Earth, jumps that correspond to permitted lines happen often enough per set of atoms that the electrons can spontaneously fall to lower energy levels. But forbidden lines can't be seen in ordinary circumstances on Earth because the density of terrestrial gas is high enough that the electrons are knocked out of the higher energy levels before they fall to the lower levels, thus preventing the transitions that give off these lines. In the corona of the Sun, however, the density is so low that the electrons can sit around on their upper energy levels long enough for them to spontaneously drop down, emitting forbidden lines.

The coronal emission lines that appear in the visible part of the spectrum, such as the red and green coronal lines, are all forbidden lines. These highly ionized species of atoms such as iron have permitted lines, too. But these permitted lines occur at very short wavelengths, too short to see in the visible. They occur in the far ultraviolet and x-ray regions, and can be studied from spacecraft.

Solar Scribblings

When a rocket went up to study the far ultraviolet solar spectrum during the eclipse of 1970, it got more than scientists expected. Not only did they find the permitted lines of iron and other elements that they expected, but they also found even stronger emission from neutral hydrogen. But the solar corona was millions of degrees, too hot for hydrogen to remain neutral. It turned out that even at those high temperatures, just a little bit of neutral hydrogen remained unionized, and that small amount was enough to scatter sunlight toward Earth to provide the observed emission. Later spacecraft were designed to make direct studies of this emission, known as Lyman-alpha, directly, to map the temperature of the corona.

Solar Neutrinos

In Chapter 3, we discussed the solar neutrino problem and its solution in terms of neutrinos changing from one type to others. The history of the solar neutrino problem is another illustration of the usefulness of the Sun as a laboratory in space.

Solar Scribblings

Trillions of solar neutrinos pass through you each second. They interact so rarely with matter that they are hard to pick up even when you are trying. Just to scare people away from our equipment at the eclipse in Australia in 2002, we hung a sign saying "Caution: Solar Neutrinos."

The neutrinos from the Sun have been monitored for about 35 years. Only in 2002 did a neutrino detector in the same Japanese mine record neutrinos emitted by power plants all over Japan. The work was used to verify that neutrinos indeed change in "flavor" after their formation.

Solar Scribblings

   When the neutrino problem first emerged in the 1960s, physicists were scornful of astronomers analyzing it. The physicists said that the astronomers just didn't understand the Sun well enough, and that they must have the temperatures in the middle of the Sun wrong. The astronomers said, in return, that they knew about the Sun and that the physicists had their basic physics of neutrinos wrong. I am glad to report that the astronomers have been vindicated and that "new physics" beyond the "standard model" of elementary particles must exist to allow neutrinos to change in "flavor." No doubt the importance of the result led to the Nobel Prize committee for Physics deciding to award half the prize for this topic.

The Least You Need to Know

- The positions of eclipses from thousands of years ago can tell us how the day has changed in length.

- Each total eclipse has its own particular story.

- The fundamental element helium is only one of many things found by studying the Sun.

- Studying the Sun helps us understand laws governing the motion of hot gas in magnetic fields.

- Only in low-density gas like the corona can we observe the forbidden lines in the spectra of elements.

- Studying the Sun can teach us basic laws of physics.

To the Ends of the Earth

In This Chapter

- ◆ The 7-minute total eclipses occur every 18 years 11⅓ days, most recently in Hawaii and Mexico

- ◆ European eclipses are popular sights

- ◆ Recent American eclipses have been only annular or partial

- ◆ African eclipses link the Sun with safaris

- ◆ The Australian outback can give clear skies

Eclipses are seen by three kinds of people:

- ◆ Scientists who travel to see them

- ◆ Tourists who specialize in traveling to eclipses

- ◆ People who are lucky enough to live along the eclipse path already

No total solar eclipses have crossed the continental United States since 1979, although there have been a couple of annular ones, plus a largely cloudy total eclipse in Hawaii.

As Long as Can Be

As we already discussed, at intervals of every 18 years 11⅓ days, the longest total eclipse in a saros interval appears. Most recently, the path of the July 11, 1991, eclipse crossed Hawaii and then traversed the Pacific Ocean to hit North America at Baja California. Totality proceeded over Mexico City and down to South America.

> **Solar Scribblings** _____
>
> The previous 7-minute eclipse to the 1991 event crossed Africa. The peak duration was in the Sahara desert north of Timbuktu, the contemporary remnant of a historically glorious city in Mali. I traveled there in advance of the eclipse to see if we could handle the logistics, but we decided eventually to observe the eclipse from Kenya, where totality was shorter but weather predictions and logistics were better.

The eclipsed Sun in the sky during the 1973 eclipse. Viewing conditions were best in Kenya, where this picture was taken, although the eclipse totality exceeded 7 minutes in the Sahara.

(Jay M. Pasachoff, Williams College Expedition)

Predictions of weather are made long in advance—not real weather predictions, but actually statistical evaluations of past cloudiness. In this case, the predictions called for a 90 percent chance of seeing the eclipse both in Hawaii, with 3 minutes of totality, and in Baja California, with almost 7 minutes of totality.

I opted for taking my team of students and colleagues to Hawaii, since we wouldn't have to go through the logistical bother of taking our equipment through customs. We were willing to sacrifice some minutes of totality for what we thought were better logistics. We were located at sea level.

Other scientists were using the major observatory on top of Mauna Kea, the volcano that was visible from our site. Some of the world's largest telescopes are there.

But a Pacific storm came up, not to mention volcanic dust blowing from the eruption of Mt. Pinatubo in the Philippines, which had started not long before. The eclipse was in the early morning in Hawaii. The day of the eclipse dawned cloudy and then cleared overhead. The question was which would happen first: the clouds dispersing or the Sun rising over the cloud rim into clear sky. Unfortunately for my site, the clouds won. We didn't see totality, which was a tremendous disappointment. But by 8 A.M., the sky had cleared and we could see the final partial phases. In the weather records of Hawaii, the day will be marked down as "clear."

On top of Mauna Kea, at 14,000 feet of altitude, the scientists did better, but barely. Clouds from the storm were rising even to the top of the mountain, and it looked as though the telescope domes would have to be shut to protect the telescopes from moisture. But the clouds held off long enough for observations to be made. The most successful were by a team led by Serge Koutchmy of the Institut d'Astrophysique in Paris. He used the Canada-France-Hawaii Telescope to get exceedingly high-resolution images of the corona, and he even detected some changes during the three minutes of totality.

Fun Sun Facts

People who travel to eclipses are often called "eclipse chasers," but I don't like that term; as we have seen, eclipses move too fast to keep up with. I've never chased an eclipse; the eclipses always catch up with me. I prefer the term umbraphile—or no special word at all.

A high-resolution image of the corona taken at the Mauna Kea Observatory during the total solar eclipse of 1991.

(Serge Koutchmy, et al.)

In Baja California, the eclipse took place almost directly overhead. The temperature was very high, but the day was largely clear. Cabo San Lucas, the resort city where most people were housed, was in the path of totality but wasn't central. People who stayed there were successful. But some of the people who made the trek to the center line were clouded out. Still, on the whole, the Baja sites were clear.

The eclipse proceeded to cross Mexico City, with its almost 20 million inhabitants, in clear weather. Perhaps as many saw this total solar eclipse as any other in history.

Rooting for Extreme Cold

The total eclipse of March 9, 1997, crossed only remote and cold terrain in northern Mongolia and Siberia. Still, the hope that we would have cold and clear weather led some of us to tackle the trip.

A special chartered train took us north from Ulan Bator, the capital of Mongolia. We went north about six hours, viewing the conical housing called yurts alongside the train tracks. But the clear, cold weather we had hoped for didn't materialize. It had actually been 50°F in Ulan Bator, and it was close to freezing at the eclipse site instead of –20°F. It actually snowed, and we tried to view the eclipse through the snow. Fortunately, the cloud cover was thin, and we got at least a glimpse of the corona, though not a good one.

Farther north, it was indeed clear and cold in Siberia. A Russian team was successful there in observing the eclipse.

Caribbean Sun

An eclipse in the Caribbean sounds good. After all, we want to see the Sun instead of clouds during the eclipse, which is the same goal that most tourists have during their vacations. The path of totality on February 26, 1998, went over the northern Galapagos Islands, the southern border of Panama, the northernmost parts of Colombia and Venezuela, and then over the islands of Aruba and Curaçao off Venezuela's coast. These islands are part of the Netherlands. Totality on parts of them exceeded three minutes. Later in the path, the French island Guadeloupe, as well Antigua and Montserrat, all in the West Indies, saw the eclipse. Totality on Guadeloupe ranged from zero through three minutes, depending on location. Many cruise ships also came into the path of totality.

A steady wind of at least 20 knots blows over Aruba, which we discovered to be a desert island. We spent a lot of our time building windbreaks, though we did get our equipment set up on the roof of a hotel.

We had largely clear weather, but about two hours before the eclipse, the sky clouded up completely! Fortunately, it then quickly cleared, and there were clouds in the sky for us during totality, but not in the direction of the Sun. Most other eclipse viewers also saw the spectacle.

The totally eclipsed Sun, viewed from Aruba during the total eclipse of 1998.

(Jay M. Pasachoff, Williams College Expedition, and Wendy Carlos)

Eclipses in Europe

The hundreds of millions of people living in Europe enjoyed the total solar eclipse of August 11, 1999. The peak of the eclipse occurred in Romania, a few hundred miles northwest of Bucharest. But the duration of totality was fairly flat over Europe, not changing very much from two minutes.

The weather forecasts indicated little chance of seeing the eclipse in Cornwall, England, where it was almost certainly going to be cloudy. Still, some diehard British tourists insisted on staying there. The weather was predicted to be spottier across Western Europe, with better weather in Eastern Europe, and that was the case. After southern England, the eclipse crossed parts of France, Belgium, Luxembourg, Germany, Austria, Hungary, Yugoslavia, Romania, and Bulgaria. The people who saw the eclipse in Germany, for example, saw it through holes in the clouds, sometimes dodging rainstorms.

I took my own scientific group of staff and students to Ramnicu Valcea, Romania, right at the point of maximum eclipse—in both duration and height in the sky. We spent about 10 days onsite getting our ton of equipment ready and aligned to track the Sun. We observed the eclipse in completely clear sky.

The eclipse proceeded over Bulgaria and to the Black Sea. Many eclipse tourists saw the event from Turkey, where it was very clear. Totality continued along its path to northern Iraq and to Iran, where the skies were also completely clear. It concluded in Pakistan and India.

At midnight following the eclipse, at our site in Romania, the bad weather reached us from the west, and a terrible windstorm tore apart the housing for our equipment, which had largely, but not entirely, been disassembled. So we had success, but only by about 12 hours.

Recent American Eclipses

Though no total eclipses have crossed any part of the U.S. mainland since 1979, two annular eclipses provided interesting views for millions.

1984 Annular Eclipse in the South

The 1984 eclipse was predicted to be annular, but just barely. The Moon was to cover 99.9 percent of the Sun, so we hoped to glimpse a bit of corona. I set up equipment in Virginia, ready for the event. But the days preceding the eclipse were cloudy and were getting worse.

Sun Safety

I lectured at a school in Williamsburg, Virginia, and was shocked and dismayed to find that the school board wasn't going to let the students out to see the eclipse, for fear of eye damage. We tried reasoning with the principal, and the teachers were on our side, but we didn't win. At least it turned cloudy there, so the students didn't miss anything they might have seen.

But had the day been clear, the students would have learned later that day that many people had seen the eclipse safely. They probably would have trusted their teachers and public officials less when, in later months or years, they issued warnings about other things—drunk driving and so on. The net loss by preventing students from seeing eclipses safely is possibly tremendous.

The night before totality, I decided that optimism is good but that realism has to win. My family and I set out to fly to New Orleans, and we intercepted friends from New

York who were about to fly to Virginia. Early in the morning, we drove up into the band of annularity, which included Picayune, Mississippi. We drove to the town square, to get the pretty steeple of a church in our views, and even found another eclipse tourist there already.

Pinhole crescents on my children during the partial phases of the annular eclipse of 1984, viewed from Picayune, Mississippi.

(Jay M. Pasachoff)

The day was perfectly clear, and the Moon indeed almost entirely covered the Sun. But even that 0.1 percent made the sky too bright to see the corona. At another location, *Sky & Telescope* editor Dennis di Cicco used an ingenious method of covering up one side of the Sun with a filter, in order to photograph a bit of innermost corona on the other side.

1994 Annular Eclipse Crossing Diagonally

The 1994 annular eclipse swept across the United States from southwest to northeast. I went to New Hampshire to see it. It was a quiet expedition—my wife, a few friends, a TV satellite truck from a Boston channel, and me. We could see the street fair going on in Boston.

We knew in advance that a big annulus would remain; the Moon was simply too small in the sky to cover the Sun. We could see the eclipse well when looking through our solar filters, but it was always too bright to look at directly or to make the sky dark.

Sun Safety

For an annular eclipse, it never becomes safe to look without special filters.

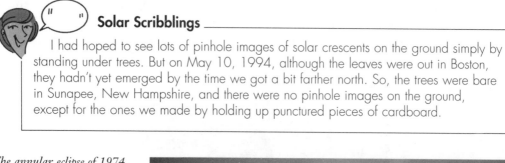

Solar Scribblings

I had hoped to see lots of pinhole images of solar crescents on the ground simply by standing under trees. But on May 10, 1994, although the leaves were out in Boston, they hadn't yet emerged by the time we got a bit farther north. So, the trees were bare in Sunapee, New Hampshire, and there were no pinhole images on the ground, except for the ones we made by holding up punctured pieces of cardboard.

The annular eclipse of 1974, viewed through a solar filter.

Partial Eclipses

Although there have been no recent American total eclipses, partial phases from two annular eclipses reached much of the United States. I viewed the annular eclipse of December 14, 2001, from Costa Rica, and the annular eclipse of June 10, 2002, from Puerto Vallarta, Mexico. In both cases, the penumbra reached the United States, making a partial eclipse for people in it.

Eclipses and Elephants

A recent pair of eclipses gave tourists views of the Sun and of fascinating animals, as well as giving professionals a chance to make scientific totality studies.

2001 in Southern Africa

On June 21, 2001, a total solar eclipse started in the Atlantic Ocean off the coast of Uruguay, reached its peak in the Atlantic west of Angola, and crossed southern Africa. Observers were stationed at various places across Africa.

Because of the decades-long civil war now settled in Angola, few tourists went there, even though it had the longest totality on land, over four minutes. The French scientist Serge Koutchmy did make his base there.

Most tourists and scientists were in Zambia or Zimbabwe. My own group was in Lusaka, the capital of Zambia, where we set up our equipment on top of a hotel and on the university grounds. The weather was gorgeous, and we even knew a day or more in advance that there would be no clouds across southern Africa. For our three minutes of totality, the corona was fairly round, showing the traits of the maximum of the solar-activity cycle.

Solar Scribblings

Tourists loved the African eclipse in part because it gave them a chance to see animals in the game parks in addition to seeing the beautiful eclipse. A general goal was to see the big five: lions, elephants, giraffes, hippos, and Cape buffalo. My team saw them all in Chobe National Park in Botswana after the eclipse.

2002 in Southern Africa

The December 4, 2002, eclipse peaked over the ocean between Africa and Australia, but nobody could reach that spot. Before noon, it crossed southern Africa, though on a slightly different track than the previous year's eclipse. In particular, it went south of Zambia, sending most tourists to Zimbabwe or South Africa.

This eclipse differed from the previous one in that it was now the rainy season in southern Africa. Therefore, the weather statistics were much less favorable: only about two-thirds chance of seeing totality. And this is what actually occurred. It was largely cloudy, but there were holes in the clouds through which many people saw totality. However, Kruger Park the big South African game preserve/National Park, was cloudy.

Outback Eclipse

Far past the peak of the December 4, 2002, eclipse, near sunset, the path of totality reached Australia. I chose to intercept it at Ceduna, a town on the coast of South

Australia. The Sun was only 9° above the horizon (less than the height of a fist held at the end of your outstretched arm), but farther inland it would be even lower. My team of 11 Williams College students and colleagues reached this location over a week in advance, after a 900-km ride west of Adelaide.

We had beautiful, clear weather for most of the time we were there, but the day before the eclipse, things turned cloudy. Eclipse day itself was totally cloudy until about two hours before the event. Then some holes appeared, and we had more hope of seeing the eclipse.

The weather statistics were about the same in Australia as they had been in Africa: about a two-thirds chance of seeing totality. It turned out that many of the 20,000 or so people in and around Ceduna—swelling the normal population of about 3,500— saw the eclipse through a big hole in the clouds. Inland, with the Sun even lower on the horizon, the weather was entirely clear and people had excellent views.

The Least You Need to Know

- ◆ Eclipses that approach seven minutes in duration are highly desirable.
- ◆ The last such long eclipse occurred in 1991 in Mexico.
- ◆ There haven't been total eclipses over the U.S. mainland since 1979.
- ◆ The annular eclipses of 1984 and 1994 were visible in the United States.
- ◆ Tourists in 2001 and 2002 combined eclipses with animal safaris.

Chapter 15

To Be in the Moon's Shadow

In This Chapter

- ◆ Africa and Turkey may be clear in 2006
- ◆ Waiting for the big one
- ◆ Across amber waves of grain
- ◆ How to view and photograph eclipses

Reminiscing about past eclipses is fun. Members of a Solar Eclipse Mailing List on the Internet send messages back and forth all the time with images and reports. But planning for future eclipses may be even more fun.

What Does the Future Bring?

In this chapter, we discuss several total eclipses:

- ◆ November 23, 2003, in Antarctica
- ◆ April 8, 2005, in the Pacific
- ◆ March 29, 2006, in Africa and Turkey, and farther into Asia
- ◆ August 1, 2008, from northern Canada across Greenland, to Siberia, Mongolia, and China
- ◆ July 22, 2009, the longest eclipse in the saros
- ◆ August 21, 2017, the next total eclipse to cross the United States

Only in Antarctica

The Earth has seven continents, but even regular travelers usually don't visit more than six. The seventh, Antarctica, is remote and cold. But it has become an important scientific site. The continent is a high plateau in its middle, and the air over the South Pole is very dry, making it a desirable place to carry out certain kinds of astronomical observations.

The closer to the South Pole you get, of course, the longer the part of the year that has the midnight sun. Because, on the other hand, periods of 24-hour darkness also last longer, most scientists carrying out observations from Antarctica have to "winter over," spending six months or so there. Tourists can take a cruise in a week or so from the southern point of Chile to a peninsula that sticks up from Antarctica.

The total solar eclipse of November 23, 2003, will be visible only from Antarctica, but not from anyplace as easily accessible as the peninsula near Chile. Making things difficult, the path of the eclipse is on the opposite side of Antarctica.

A Russian icebreaker has been converted to a passenger cruise ship and regularly takes Antarctic cruises, usually out of Australia. It is scheduled for 28 days at sea to reach the location where the eclipse will be visible. I hear the Southern Ocean can be very rough. But I have a friend who has made over 10 Antarctic cruises because she thinks they are so wonderful. Whether the eclipse tourists will view the eclipse from the ship itself or from the adjacent ice is yet to be determined. The ship is to leave Port Elizabeth, South Africa; visit some islands en route; and wind up after the eclipse in Hobart, Tasmania, Australia.

The eclipse path doesn't cross any established outposts. It doesn't go too far from Russia's Mirny Station.

A couple of possibilities are arising for airplane tourists. One expedition will include several airplanes, which will land on the ice and then make short tourist trips around.

Much easier on the stomach than the 28-day-at-sea cruise will be a 1-day aircraft flights out of Perth, Australia, and out of Punta Arenas, Chile. Those who want to see the eclipse have to purchase the seats next to the windows on one side of the plane. The view from a plane is not as glorious as the view from the ground, but it will still be spectacular. And the planes can get near to the maximum point, where the eclipse lasts 1 minute and 57 seconds, and prolong the eclipse by 15 additional seconds or so.

Partial phases will be visible from the southern tip of Chile and Argentina, and also from Australia and New Zealand.

No Land—Maybe Just a Little

The eclipse of April 8, 2005, will be annular. The Earth won't be far past the tip of the Moon's shadow cone, so the track will be very narrow. The eclipse starts in the ocean southeast of New Zealand. At its end, it will intersect Central America. Even the maximum duration of annularity is only 42 seconds. But even more interesting, the eclipse will be total for a second or so in the middle of the Pacific Ocean.

We have to hope that there are enough umbraphiles to support a cruise ship going way out there for the event. At this writing, it is too early to know.

Turkey?

The next major total solar eclipse won't be until March 29, 2006, so there will clearly be a lot of eclipse tourists wanting to see it. The eclipse will start at the easternmost tip of Brazil and then will cross the Atlantic Ocean. It will hit West Africa in Ghana, Togo, Benin, and northern Nigeria. It then will go over Niger, east of the desert city of Agadez at the foot of the Sahara close to where the 1973 long eclipse crossed. Then the path will cross Chad and Libya and barely touch northwestern Egypt. Totality will peak at 4 minutes 7 seconds close to the border between Chad and Libya.

Off the center line, with diminished totality, it will cross the Greek island of Castellorizo. Perhaps the major viewing sites will be in Turkey; the path goes through Turkey's center, east of Ankara.

Then the eclipse will go up to countries that were formerly in the Soviet Union, crossing the top of the Caspian Sea into Kazakhstan. The eclipse will wind up in northeast China.

Partial phases of this eclipse will be visible from all of Europe, from the western part of Asia (up to the middle of Mongolia and to Bangladesh), and almost all of Africa, missing only the southernmost parts (which have been blessed with their own total eclipses, or at least partial phases, in recent years).

Way Up North, Plus

On August 1, 2008, in the prime of summer, a total solar eclipse will start in the islands of northernmost central Canada (Victoria, Prince of Wales, Somerset, Devon, and Ellesmere Islands) and then proceed over the northwestern tip of Greenland. It will cross northern oceans, where it will hit its peak duration of 4 minutes and 27 seconds. The eclipse then will hit land in northern Russia. It will cross western Mongolia and wind up in central China. I daresay that central Russia or China will be the places to view it from. Around sunset, it will reach near to Shanghai.

As Good as It Gets

As soon as one very long eclipse occurs, people start looking one saros later. The long eclipse of July 11, 1991, viewed especially in Hawaii and Mexico, will be followed 18 years 11⅓ days later on July 22, 2009. The eclipse's peak will be "only" 6 minutes 39 seconds, and it will be shorter from all viewing spots on land; the eclipses in this saros are now getting shorter than the 7 minutes they had been.

The eclipse will start at sunrise at India's west coast and will cross India and the eastern tip of Nepal. Totality also will cover parts of Bangladesh, Sikkim, Bhutan, northern-most Myanmar, and China. The eclipse will cross Shanghai, which will have over five minutes of totality, and then go out to the Pacific Ocean, passing not far south of Okinawa in southernmost Japan and crossing some of the smaller Ryukyu Islands. The peak totality duration occurs farther along, in the Pacific.

The partial phases of this 2009 eclipse will be visible from not only India and the other countries mentioned, but also Russia, southeast Asia north of the middle of Sumatra and Borneo, and Papua New Guinea. Even the northern tip of Australia's Cape York will get a slight partial eclipse.

Home Again

Skipping ahead, let's discuss the next total solar eclipse that will be visible in the United States. On August 21, 2017 (one saros later than the eclipse that crossed Europe in 1999), totality will start in the north Pacific Ocean. It will hit land at northern Washing-ton and cross central Idaho, mid-Colorado, mid-Nebraska, northern Missouri, southern-most Illinois, western Kentucky, middle Tennessee, southwestern North Carolina, and the heart of South Carolina. Maximum totality of 2 minutes 40 seconds will occur in Kentucky; the eclipse will exceed 2 minutes all along the center line in the United States.

All the United States will see a partial eclipse, ranging from over 60 percent in Southern California, to 80 percent in Florida, to 75 percent in New York City, to just under 60 percent in northernmost Maine. Hawaii will have 30 percent to 40 percent, and Alaska will have 35 percent to 70 percent. Northern South America, from the middle of Peru and Brazil upward, and all of Central America, Mexico, and Canada will also have a partial eclipse. The partial zone will include Greenland and Iceland, and will even reach to northwestern Scotland, which saw a partial or annular eclipse in 2003.

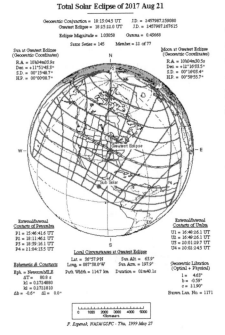

Total Solar Eclipse of 2017 Aug 21

Geocentric Conjunction = 18:13:04.5 UT   J.D. = 2457987.259080
Greatest Eclipse = 18:25:32.0 UT   J.D. = 2457987.267615
Eclipse Magnitude = 1.03058   Gamma = 0.43668
Saros Series = 145   Member = 22 of 77

Sun at Greatest Eclipse
(Geocentric Coordinates)
R.A. = 10h04m03.9s
Dec. = +11°51'43.3"
S.D. = 00°15'48.7"
H.P. = 00°00'08.7"

Moon at Greatest Eclipse
(Geocentric Coordinates)
R.A. = 10h04m30.5s
Dec. = +12°16'33.5"
S.D. = 00°16'03.4"
H.P. = 00°58'55.7"

External/Internal
Contacts of Penumbra
P1 = 15:46:41.6 UT
P2 = 18:11:46.2 UT
P3 = 18:39:16.1 UT
P4 = 21:04:13.5 UT

External/Internal
Contacts of Umbra
U1 = 16:48:26.1 UT
U2 = 16:49:26.1 UT
U3 = 20:01:29.7 UT
U4 = 20:02:24.5 UT

Local Circumstances at Greatest Eclipse
Lat. = 36°57.9'N   Sun Alt. = 63.9°
Long. = 087°38.0'W   Sun Azm. = 197.9°
Path Width = 114.7 km   Duration = 02m40.1s

Ephemeris & Constants
Eph. = NewcombMLE
ΔT = 80.8 s
k1 = 0.2724880
k2 = 0.2723810
Δb = -0.6"   Δl = 0.0"

Geocentric Libration
(Optical + Physical)
l = 4.63°
b = -0.59°
c = 21.90°
Brown Lun. No. = 1171

0   1000   2000   3000   4000   5000
Kilometers

F. Espenak, NASA/GSFC - Thu, 1999 May 27

The path of totality at the August 21, 2017, total solar eclipse across the United States, with partial phases to the sides. For the partial phases, numbers show the percentage of the solar diameter that is covered.

(F. Espenak, NASA's Goddard Space Flight Center)

Once More, with Passion

After waiting nearly 40 years, from 1979 to 2017, for a total eclipse, people in the continental United States will have to wait fewer than 7 more for the following one. On April 8, 2024, a total solar eclipse will start over the Pacific and reach land in western Mexico. It will hit the United States at Texas and will cross eastern Texas, eastern Oklahoma, southeastern Missouri, southern Illinois, most of Indiana, and northwestern Ohio. It will then clip northwestern Pennsylvania, northern New York, Maine, Vermont, and New Hampshire as it moves across northern Maine into eastern Canada at New Brunswick.

The maximum duration of the eclipse in Mexico, will be 4 minutes 28 seconds, though U.S. durations won't be much shorter. All of the United States excluding Alaska, as well as all of Canada, Mexico, the Caribbean, and Central America, will have a partial eclipse.

The Dregs

Some eclipse fans who would travel halfway around the globe for a total eclipse say they wouldn't even go out their doors for a partial one. But I don't mind getting a glimpse (through a filter for the partial ones) of them all.

October 14, 2004

This eclipse is never total or annular on Earth, but up to 93 percent of the Sun will be covered. The partial phases will be visible in Japan, northeastern Siberia, western Alaska, and the Aleutian Islands. The zone of eclipse extends barely to Hawaii.

April 8, 2005

This one is really the dregs in the United States, since you could go see the annular eclipse in Costa Rica, Panama, Colombia, and Venezuela. In any case, partial phases will be visible from about half the United States. Everyone south of a line extending roughly from San Diego to Denver, to Chicago, to Philadelphia will see a partial eclipse.

Eclipse Viewing and Photography

An eclipse is so glorious to see, with the sky changing dramatically around you even as the diamond rings and corona become visible overhead, that people often say that they wish they weren't taking photographs and could just relax and look at it. So, my first recommendation to eclipse tourists is just to relax and look around. After all, many others of us are taking photographs, and you can easily arrange to get copies. But if you want to take your own photographs, here are some hints.

Sun Safety

Remember that as long as the Sun isn't totally eclipsed, you can't look at it without a special filter. During the partial phases, you can use a filter, project a solar image onto a screen using a telescope or binoculars (without looking up at the Sun through them), or make a simple pinhole camera. But during totality, none of those methods will allow you to see the eclipse. You must look at the totality directly.

Still Photography with Film

The size of the Sun and the corona on 35-mm film with an ordinary camera lens is pretty small—only a couple of millimeters across. Such a photograph isn't usually interesting, unless you can put the Sun at the top of your frame and put some foreground objects in the rest of the frame. It is best to use a tripod and a cable release so that your camera doesn't shake.

Eclipse photographs centered on the Sun start to get interesting with telephoto lenses about 300 mm or longer. Then the Sun and the corona take up a substantial part of the frame. I recommend taking a few photographs during the diamond-ring effect and then, during totality, bracketing widely. That is, take one frame at every conceivable exposure. With my f/8 500-mm telephoto lens, I start at $\frac{1}{125}$ s and take a photograph at each click-stop of shutter speed: $\frac{1}{60}$, $\frac{1}{30}$, $\frac{1}{15}$, $\frac{1}{8}$, $\frac{1}{4}$, $\frac{1}{2}$, and 1 second, using film between ISO 100 and ISO 400. Using faster film than that usually hurts more than it helps because the grain is worse for the faster films.

The following table lists the field of view for various lenses when used with 35-mm cameras. The Sun and the Moon are about 0.5° across; allowing for 1 solar radius of corona on each side of the Sun gives a desired field at least 1.5° across.

Field of View

Focal Length (mm)	Diagonal Size	Field of View
20 mm	125°	100° × 70°
28 mm	85°	73° × 50°
50 mm	50°	41° × 28°
200 mm	12°	10° × 7°
500 mm	5°	4° × 3°
1,000 mm	2.5°	2° × 1.4°

The Solar Scoop

The size of the Sun on film depends on the focal length of your lens. It is about the same size on the film in millimeters as the focal length of your lens divided by 109—and it is good enough to just divide by 100, which is easy to do in your head. So, a 1,000-mm telephoto lens gives an image about 10 mm across. A 50-mm lens, which is a normal lens for a camera, gives an image only 0.5 mm across, too small to be attractive by itself.

The corona falls off in brightness so rapidly—a factor of 1,000 in the first 2 solar radii outside the edge of the Sun—that a mere difference of a factor of 2 or 4 in exposure time shows you a little more corona or a little less corona but doesn't ruin your picture. So, in some sense, eclipse photography is easy, since every exposure is good. The key things to watch are the focus and the steadiness.

You should have a solar filter on your lens before totality so that you can focus carefully through it, whether or not you take photos of the partial phases. It is particularly useful to focus on the sharp horns of the solar crescent. Note that when you look through an ordinary filter, which is made of flat glass, the focus doesn't change when you take it away.

> **Solar Scribblings**
>
> If you are using a point-and-shoot camera, put a piece of black tape over your flash. It may try to go off automatically, and you don't want to ruin photographs your neighbors may be taking or damage their eyes' dark adaptation.

Don't leave your camera on any automatic setting. In particular, if you leave the focus on automatic, the camera may well "hunt" through the whole eclipse—that is, go back and forth through the focus, never finding the right focus—and never take any pictures at all!

Exposure Times for Prominences

Focal Ratio	ISO 100	ISO 400
f/2	$\frac{1}{1,000}$	$\frac{1}{4,000}$
f/4	$\frac{1}{250}$	$\frac{1}{1,000}$
f/8	$\frac{1}{60}$	$\frac{1}{250}$

Exposure Times for Diamond Ring and Inner Corona

Focal Ratio	ISO 100	ISO 400
f/2	$\frac{1}{125}$	$\frac{1}{500}$
f/4	$\frac{1}{30}$	$\frac{1}{125}$
f/8	$\frac{1}{8}$	$\frac{1}{30}$

Exposure Times for Middle Corona

Focal Ratio	ISO 100	ISO 400
f/2	$\frac{1}{30}$	$\frac{1}{125}$
f/4	$\frac{1}{8}$	$\frac{1}{30}$
f/8	$\frac{1}{2}$	$\frac{1}{8}$

Still Photography with a Digital Camera

Digital cameras work quite well at eclipses. Make sure that you don't have to download images during totality. To guarantee your not having to do so, empty your storage medium before the eclipse or at least 10 minutes before totality.

Again, if you have only a normal lens, you are better off taking photographs that have the eclipse near the top of the frame and other features in the foreground. Be sure to take off all automatic settings. The auto focus may hunt, never finding a good focus because the image of the corona in the sky is so small. And the automatic exposure may do one or both of the following:

♦ Overexpose the corona, since it is measuring largely black sky and trying to bring it to a gray level

♦ Stay open much too long, taking up much of the eclipse with a single frame

The more megapixels you have, the better, to give good resolution on your images. Cameras with at least 5 megapixels (that is, millions of pixels) match the resolution of 35-mm film and enable you to make enlargements of your images. (These cameras have about $2,000 \times 2,500$ picture elements, making 5 million in all.)

Fun Sun Facts

The era of film is ending, both for home photography and astronomical photography. Both professionals and amateurs are now using charge-coupled devices (CCDs) instead of film. These devices are much more sensitive than film. The intensity of light hitting each individual picture element ("pixel") is read off into a computer. A CCD with 3,000 rows of 2,000 pixels in a row has $3,000 \times 2,000 = 6$ million pixels, or 6 megapixels.

Take off all automatic functions. Choose one exposure time for the several exposures during the diamond-ring phase (you don't want to be changing things during this brief interval), and then take exposures at many different values during totality. Be sure that the camera is mounted steadily on a tripod. And use a cable release to minimize shake.

Eclipse Videography

Video cameras have lower resolution than still cameras, but you can still make some nice movies of a total eclipse. The partial phases change so slowly—taking over an hour—that it is boring to just let the video camera run during those phases. Many cameras have ways of taking a burst of a few seconds; do so every five minutes or so. Then let the camera run starting a minute or two before totality.

The DV cameras take digital video, as opposed to earlier cameras such as Hi-8 that take video in analog fashion. Digital video has higher resolution.

Image Size on a 13-Inch Diagonal Television Screen (Camcorder Lens)

Focal Length	Size of Sun	Size of Sun
	(½-inch CCD) mm (in.)	*(⅔-inch CCD) mm (in.)*
50 mm	20 (0.8)	17 (0.7)
100 mm	40 (1.6)	34 (1.3)
200 mm	80 (3.2)	68 (2.7)
500 mm	200 (7.9)	170 (6.7)
1,000 mm	400 (15.8)	340 (13.5)

If you want to take a close-up view using a telephoto lens or use your camera on maximum optical zoom, it is best for you to obtain a mount that tracks the Sun. After all, the Sun moves its own diameter across the sky in only about two minutes. So during a reasonably long eclipse, it could just move out of your field of view if you are using an ordinary tripod. Mounts like that are available at stores (brick-and-mortar or online) that sell telescopes.

Wide-angle movies during an eclipse can be very nice because they show not only the corona in the sky, but also the changes in the atmosphere below. Be sure to turn off the automatic exposure function, or it will compensate perhaps too much for the darkening sky effect that you want to record.

Be sure to turn off the automatic focus and to set your focus for infinity (or focus in advance on the Sun through a solar filter or on some very distant object). Otherwise, your camera may go back and forth through the focus throughout totality and never get a clear image.

Many things can go wrong when you are trying to photograph an eclipse. Be sure that you have checked the following:

- Your battery is charged.

- There is film, tape, or a digital card in the camera, as appropriate.

- Your filter or lens cap is off for totality.

- You can get your filter off easily without shaking the camera as totality starts.

- The Sun doesn't drift out of the field of view during totality.

- You don't make the camera vibrate in your exuberance to snap the shutter.

- The REC symbol really shows, if you are using a video camera, so that you know that you are recording.

- Nobody trips on the power cord if you are plugged into an electrical outlet.

- Nobody bumps into your tripod.

The Least You Need to Know

- Interesting eclipses can be seen about every 18 months somewhere in the world.

- The next eclipses that will be subject to the most eclipse tourism will be those in Turkey in 2006 and China in 2009.

- The next total eclipses visible in the United States will be in 2017 and 2024.

- It isn't hard to photograph an eclipse, but you must be careful in various ways to observe safely and get a good image.

Chapter 16

Venus Tries to Cover Immodestly

In This Chapter

- ◆ Measuring the solar system is hard
- ◆ Transits of Venus took us to the south seas
- ◆ Transits of Mercury are practice
- ◆ We get a chance soon

Total solar eclipses, when the Moon goes in front of the Sun and blocks its light entirely, are rare. Yet if you travel, you can see one every year and a half or so. One astronomical phenomenon comes around with a much longer interval of over 100 years. This event is a transit of Venus, a passage of Venus across the face of the Sun. Nobody alive has ever seen one, yet two are coming up in the next decade. What an opportunity!

A Blot on the Face of the Sun

On November 24, 1639 O.S., 20-year-old Englishman Jeremiah Horrocks searched for a transit of Venus. Kepler, who had predicted the 1631 transit, didn't predict this one, but Horrocks had re-evaluated Kepler's calculations. He started observing early, in case the predictions were off. When some

Sun Safety

Horrocks was observing the Sun by projecting its light onto a screen. Remember that you should never stare at the Sun directly without special filters. Special filters or projection are needed to observe transits of Venus or of any other object that is not fully blocking the everyday solar surface.

clouds cleared up, he clearly saw a circular black spot silhouetted against the Sun. He had alerted a friend some miles away, who got a glimpse of this circular black shape through a hole in the clouds that briefly opened at his site. Could it have been a sunspot that Horrocks saw?

But what he had seen was even rarer. The black spot was Venus silhouetted against the Sun. Though Venus is larger than the Moon, it is much farther away and, therefore, takes up a much smaller angle on the sky. So rather than blocking out the Sun, it merely shows as a spot only about 3 percent the Sun's diameter. Unless you were observing the Sun, you wouldn't notice it.

The Solar Scoop

Dates given in Old Style (O.S.) are from before the calendar reform that took 10 days off the schedule, making the Gregorian calendar we use today. On the European continent, the change was made in 1582. But in England, where Horrocks was observing, and its American colonies, the change didn't take place until 1752. So the date of the 1639 transit was recorded differently depending on location.

After Edmond Halley figured out that a certain apparent set of comets was really one single comet returning periodically—an object that we now call by his name—he turned his attention to other topics. Earlier, at the age of 21, he had observed a transit of Mercury while on the isolated mid-Atlantic island of St. Helena to make a southern star catalogue. Much later, in 1716, Halley figured out that transits of Venus could be the way to solve one of the major problems of astronomy: How big is the solar system? At the time, the average distance of the Sun from Earth, which is known as the astronomical unit, was known to an accuracy of only about 20 percent—that is, it was somewhere between 135 million km and 165 million km (about 80 million miles and 110 million miles). In 1720, Halley became Astronomer Royal.

In Chapter 4, you learned how Johannes Kepler figured out the basic laws of how planets moved around the Sun. His first two laws were part of his book *Astronomia Nova* (*The New Astronomy*), published in 1609. His first law states that the planets orbit the Sun in ellipses, with the Sun at one focus of each ellipse. His second law states that if you draw a line connecting a planet with the Sun, in each time interval of a certain duration, that line sweeps out an area that is always the same. This law implies that when a planet is relatively close to the Sun, it moves faster in its orbit, since the triangle that the line sweeps out has to be wide and squat compared with the narrow, tall tri-angle swept out at other times. In 1618, in Kepler's book *Harmonices Mundi* (*On the Harmony of the World*), he advanced his third law: The period of a planet's orbit cubed is equal to the size of its orbit squared, when given in units of Earth's orbit.

Sun Words

A **transit** is a passage of an astronomical body in front of another. An object passing in front of the Sun is called a transit of that object. A solar eclipse, indeed, is a type of transit, one involving the Moon. Only two other objects of substantial size in our solar system transit the Sun: the planets Mercury and Venus, whose orbits are within Earth's. Astronomers can now pick up exoplanets—planets that orbit other stars—transiting in front of their stars.

All Kepler's relationships were relative: Jupiter's orbit is 5.2 times bigger than Earth's, and its period is 10 times longer. But nowhere in Kepler's laws does it say how big Earth's orbit is or how big Jupiter's orbit is in actual units of length, such as kilometers or miles.

Halley had the brilliant idea of using a transit of Venus to compute the actual size of the solar system. By the time he thought of it, he knew that the next transits of Venus would be only a few decades away, in 1761 and 1769. Halley realized that if people looked at the transit from different locations on Earth, the line from each telescope to Venus and onward would hit the Sun at a slightly different place. The farther Venus is from the Earth, the smaller the deviation of the position at the Sun would be. By timing accurately how long Venus would take to cross the Sun, scientists could later calculate how far away Venus was.

With the prospect in mind of solving the problem of the size of the solar system, many nations equipped and sent out scientific expeditions to the 1761 and 1769 transits. To apply Halley's method, it was important to have the expeditions spaced as widely as possible across the face of the Earth.

Solar Scribblings

English surveyors Charles Mason and Jeremiah Dixon were sent by their government to Sumatra to observe the 1761 transit, but their ship was attacked by the French before they got too far. After returning home with their dead and wounded and repairing the ship, they made it as far as Cape Town, South Africa. Their results were so good that they were next dispatched to the American colonies to survey the disputed Pennsylvania-Maryland border. The result was the Mason-Dixon line.

For the 1769 transit, the British Admiralty, which administered the world's greatest sea power, appointed a young lieutenant, James Cook, as captain of its ship, the *Endeavour*. Captain Cook was sent to Tahiti, in the South Pacific, to observe the transit; he took with him astronomer Charles Green. Of course, Captain Cook became one of the most famous sailors of all time from this expedition, which went on to sight Australia and the islands of New Zealand for the first time for Europeans. There had been rumors of a "Great Southern Continent," and Captain Cook went on to find out what he could about it. As a result, I like to consider the European colonization of Australia and New Zealand a spin-off of astronomy.

The results of the various expeditions gave distances for the astronomical unit that ranged from 93 million to 97 million miles, an uncertainty of about 2 percent around a central value of 95 million miles. We know now that the actual value— 92,957,000 miles— is at the bottom of the range.

Good Things Come in Pairs

Transits of Venus may be very rare, but they come in pairs. Each pair is separated by only eight years. But then there is a gap of over 100 years until the next pair.

Just as eclipses of the Sun occur only when the Sun, the Moon, and Earth are in line, transits of Venus occur only when the Sun, Venus, and Earth are in line. When that happens depends on the orientations and tilts of the planets' orbits around the Sun. Note from the following list that these nodes in the orbit occur only in June or December. The gap alternates between about 105 years and about 122 years.

Transits of Venus:

- 1631: December 7 (unobserved)

  1639: December 4 (November 24 o.s.)

- 1761: June 6 (after a gap of nearly 122 years)

  1769: June 3–4

- 1874: December 8–9 (after a gap of 105 years)

  1882: December 6

- 2004: June 8 (after a gap of nearly 122 years)

  2012: June 5–6

- 2117: December 11 (after a gap of 105 years)

  2125: December 8

The Black-Drop Effect

Timing the transit required accurate clocks. The best clocks of that time used pendulums to beat steady time and had to be set up on firm surfaces, not on the desks of ships on the rolling seas. In her best-selling book *Longitude*, Dava Sobel beautifully described the trials that her hero went through to get his spring-wound clocks accepted.

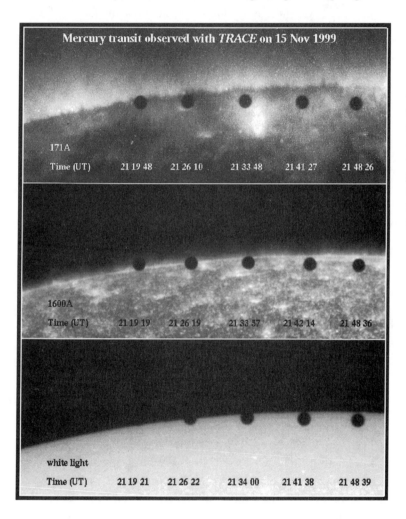

The 1999 transit of Mercury observed with the Transition Region and Coronal Explorer (TRACE) spacecraft. The image at 171 A is from million-degree iron (8-times ionized); the image at 1600 A is taken of continuous radiation from chromospheric heights, and the white light image shows Mercury silhouetted against the photosphere.

But even with a decent clock, a completely unexpected problem appeared: It was not possible to accurately measure the instant when Venus became fully silhouetted against the Sun. It had been thought that you would merely see Venus's disk touching the edge of the Sun. But instead, as Venus moved entirely in the Sun, a black band—curved on each side—seemed to join Venus with the outside sky. The band seemed

Sun Words _____

The **black-drop effect** is the appearance of a dark band joining Venus's silhouette and the sky, preventing accurate timing of the transit.

like a black drop of fluid coming in from the edge of the Sun, and the effect became called the _black-drop effect_. The band pulled out, like taffy, and eventually snapped, showing Venus fully silhouetted. But 10 seconds or so of uncertainty had ensued. That uncertainty in time by an unexpected factor of 5 translated to an uncertainty increased by the same factor of 5 in the distance of Venus from Earth.

Lengthy Expeditions

In the twenty-first century, you can fly off to Antarctica, see a solar eclipse, and be home in a week. In the early twentieth century, an eclipse expedition to Russia in 1936 involved many weeks in a chartered train to take the equipment and crew out to Siberia. But these travails pale in insignificance next to the effort necessary for eighteenth-century expeditions to transits of Venus.

One of the leading scientists in the field was Guillaume Joseph Hyacinthe Jean Baptiste le Gentil de la Galasière of France. Le Gentil had carried out some of the fundamental analyses of the theory, extending Halley's work. For the transit of 1761, he set out in a French vessel to Pondicherry, a French possession in India. The French were at war with the English, who controlled the seas in general, in addition to making a specific siege of Pondicherry. He managed to obtain a letter guaranteeing him safe passage. Weeks later, he was prevented from landing in India because of the fighting. He could observe the transit, which took place in clear skies, only from the ship. The rocking of the ship meant that his clock wasn't accurate. Thus, he could not make the measurements that he needed.

Le Gentil decided to wait for the transit of 1769, which was, after all, "only" eight years later. This time, he was almost arrested as a spy at his choice site in Manila, but he managed to get to Pondicherry a year in advance of the event to set up his equipment. Hours before the event, however, the sky clouded up, and he was again foiled. His trip home rivaled Ulysses's in difficulty. Le Gentil fell sick, was hospitalized for some months, and was shipwrecked. It took years for him to get home, making his total absence 11 years and 6 months. By the time he returned, he had been declared dead and his estate had been divided up among his heirs. His seat in the French Academy, the high academic honor, had been given away. His personal story eventually ended happily, after years of misery.

(Wendy Carlos, Jay M. Pasachoff, Stephen Martin, and Daniel B. Seaton/Williams College Expedition)

The 1999 total eclipse of the Sun, in a composite made from electronic and photographic images to bring out the inner coronal structure.

(Jay M. Pasachoff and Wendy Carlos; center: SOHO/EIT Team)

A false-color space image of the corona over the face of the Sun has been inserted at the center of the image.

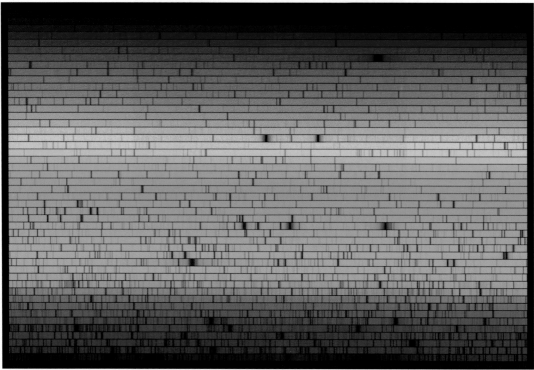

(N. A. Sharp, NOAO/NSO/Kitt Peak FTS/AURA/NSF)

The solar spectrum, cut into a set of strips. The left side of each strip, except for the first, would really be joined to the right side of the one above it.

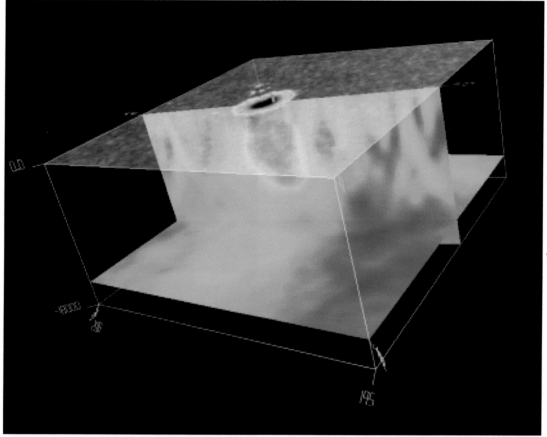

(MDI on NASA/ESA Solar and Heliospheric Observatory)

The speed of sound below a sunspot, measured by the Michelson Doppler Imager experiment on the Solar and Heliospheric Observatory. These results of helioseismology show the sunspot on the top, with its dark umbra surrounded by a lighter penumbra. Below it, we see to a depth of 24,000 km (15,000 miles). Faster sound speed shows as red and slower sound speed as blue.

The images on this page and the next show the Sun in different wavelengths or different modes, all on the same day, February 8, 2001.

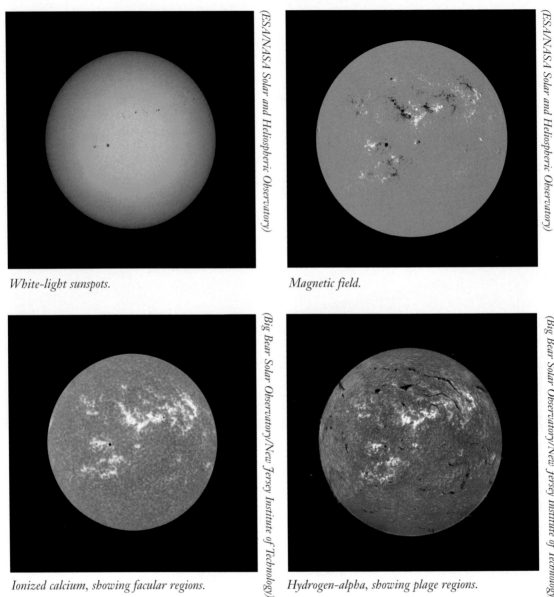

(ESA/NASA Solar and Heliospheric Observatory)

(ESA/NASA Solar and Heliospheric Observatory)

White-light sunspots.

Magnetic field.

(Big Bear Solar Observatory/New Jersey Institute of Technology)

(Big Bear Solar Observatory/New Jersey Institute of Technology)

Ionized calcium, showing facular regions.

Hydrogen-alpha, showing plage regions.

(ESA/NASA Solar and Heliospheric Observatory)

Helium in the extreme-ultraviolet from SOHO, showing the chromosphere at about 50,000°C.

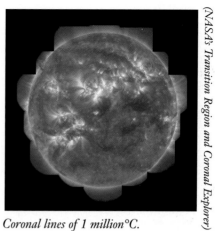

(NASA's Transition Region and Coronal Explorer)

Coronal lines of 1 million°C.

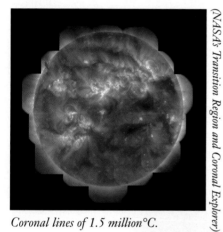

(NASA's Transition Region and Coronal Explorer)

Coronal lines of 1.5 million°C.

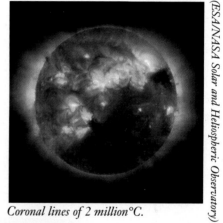

(ESA/NASA Solar and Heliospheric Observatory)

Coronal lines of 2 million°C.

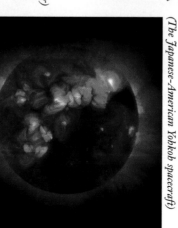

(The Japanese-American Yohkoh spacecraft)

X-ray image from Yohkoh at 3 to 5 million°C.

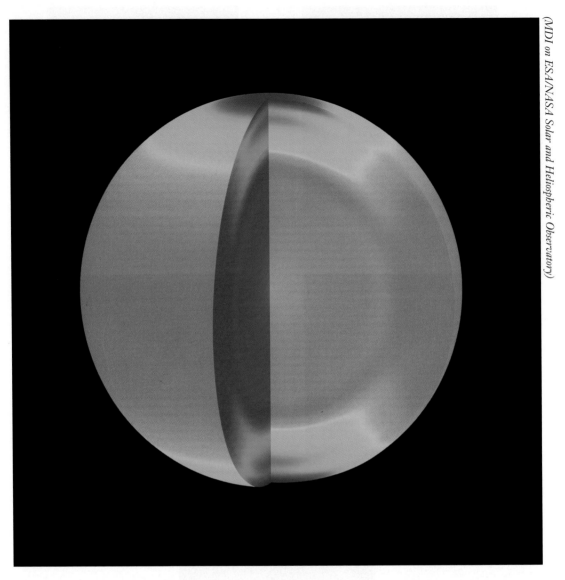

(MDI on ESA/NASA Solar and Heliospheric Observatory)

The speed of sound, measured with helioseismology from SOHO. We see both the equator-to-pole variation and the surface-to-core variation. Red indicates rotation faster than average and blue indicates slower than average. Helioseismology reveals the internal structure.

(Swedish Solar Telescope, Royal Swedish Academy of Sciences, with TRACE image, Lockheed Martin Advanced Technology Center; at top)

The Sun, at four different levels in the atmosphere. We see sunspots at the bottom, surrounded by granulation. Above it, we see chromospheric radiation in the light of the calcium K line at a temperature of about 10,000°C. Above that, we see the Sun in hydrogen light, in which features tend to trace the Sun's magnetic field. The top image shows million-degree gas from the corona.

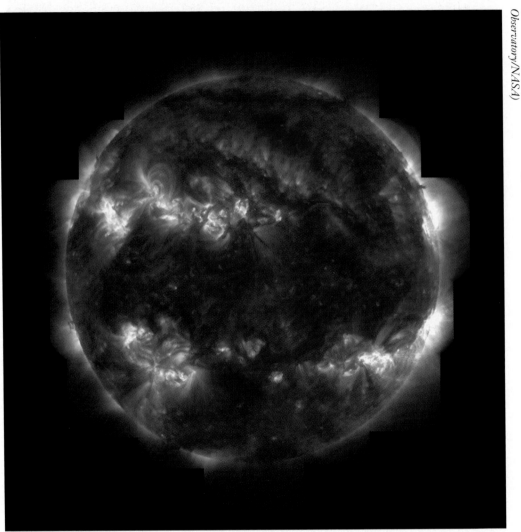

A false-color image combining the three coronal wavelengths observed by the Transition Region and Explorer (TRACE) spacecraft.

The expeditions may have been difficult, but the data were valuable. The results published two years after the second transit of the pair found 153 ± 1 million km for the size of the astronomical unit. A later analysis, by the famed nineteenth-century American astronomer Simon Newcomb, used better methods with the same data to find 149.59 million kilometers. Today's value, measured to within meters using radar and adopted as the official value of the astronomical unit by the International Astronomical Union, is 149,597,870 km.

Solar Scribblings

The story of Le Gentil's expeditions has been told in a 1990s play, *The Transit of Venus*, by the Canadian playwright Maureen Hunter.

American Expeditions

Out of dozens of expeditions from countries around the world, eight American expeditions, funded by Congress, went to the 1874 transit of Venus, some in the United States and others abroad. Observation sites for the event were all over the world, as far south as Tasmania and New Zealand and as far north as Siberia. But there were instrumental and weather problems at most of the sites, and the results were not considered good. Eight more expeditions were funded for the 1882 event. Though it has been said that no values for the size of the solar system resulted, there actually were results, though only after years of data reduction. When put together with other methods available at the time of determining the distance to the Sun, the results were not heavily weighted. By this time, observations from distant locations of Mars and of asteroids gave reasonably good values for the astronomical unit.

One of the best-known American observers of the 1882 transit was Maria Mitchell, who had gained fame as a girl when she discovered a comet, for which she had won a gold medal from the King of Denmark. At the time of the transit, Mitchell was a professor of astronomy at Vassar College in Poughkeepsie, New York. She and her students observed the transit from the College Observatory. The photographs Mitchell took of the transits of 1874 and 1882 are an important part of the early history of astronomical photography.

Solar Scribblings

Mary Lyon, founder of the Mt. Holyoke Seminary for Women (now Mount Holyoke College) in South Hadley, Massachusetts, sent two teachers to South Africa in the mid-nineteenth century to found a seminary there. Mt. Holyoke also transferred its 1853 telescope to the South African seminary in time for it to be used to observe the 1882 transit of Venus.

The 1882 transit of Venus, photographed by Maria Mitchell and her students at Vassar College.

(Special Collections, Vassar College Libraries)

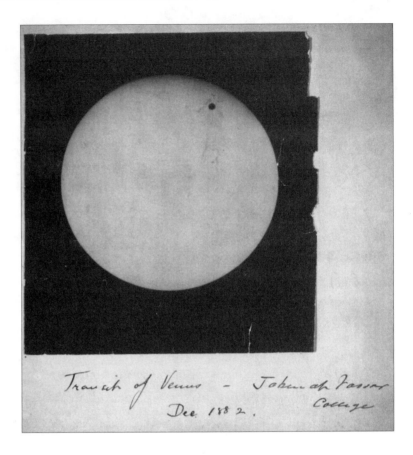

Not Venus's Atmosphere

When Venus is near to the Sun in the sky but is not silhouetted against it, a bright ring around Venus has been observed. The effect is that of Venus's atmosphere refracting (bending) sunlight toward us. In the second half of the twentieth century, Venus's atmosphere was explored by spacecraft that flew by, orbited, or even landed on the planet. Venus has a thick, dense, atmosphere, with cloud cover so heavy that we cannot see through to the exceedingly hot and unpleasant surface.

Because Venus's atmosphere is so well known, many, if not most, people have assumed that the black-drop effect is caused by Venus's atmosphere. But the atmosphere is not even close to thick enough to create an effect the size of the black drop. Still, many books and articles persist with this incorrect explanation.

Mercury to the Rescue

Though at the time of a transit Venus is as close as a planet ever can be to Earth, the planet Mercury also transits the Sun and can be used to measure the Sun's distance. Mercury is smaller than Venus and also over twice as far away, so its silhouette against the Sun is even smaller than Venus's. Mercury is only about half of 1 percent of the size of the Sun in angle in the sky: 10 arc seconds compared with 30 arc minutes (30 × 60 = 1,800 arc seconds).

A transit of Mercury was observed historically even before a transit of Venus. French astronomer Pierre Gassendi saw the Mercury transit of November 6, 1631, which had been predicted by Kepler on the basis of the *Rudolphine Tables* he had compiled.

Sun Safety

As with transits of Venus, you must use special filters or projection methods to view transits of Mercury, since the everyday surface of the Sun is visible throughout.

Mercury's transits are at least 10 times more common than Venus's. There are about a dozen per century, but because Mercury's orbit is tilted 7° with respect to Earth's, only about one out of two dozen of Mercury's orbits lead to a transit. Each transit lasts about five hours. The beginning of the November 8, 2006, transit of Mercury will be visible from all of the western hemisphere.

The next table illustrates how transits of Mercury, like transits of Venus, occur in pairs. In both cases, the pairs represent opposite nodes of the orbits. Transits of Mercury occur only in May and November.

Transits of Mercury:

◆ 1993: November 6 (corona only)

  1999: November 15

◆ 2003: May 7

  2006: November 8–9

◆ 2016: May 9

  2019: November 11

◆ 2032: November 13

  2039: November 7

Solar Scribblings

The beginning of the November 8, 2006, transit will be visible in the afternoon from all of North and South America. Hawaii, but not the rest of the United States, will be well placed for midtransit. The end of the transit will be visible from all of Australia, but only from the extreme east coast of Asia.

The transit will last from 19:13 U.T. to 00:11 U.T., with midtransit at 21:42 U.T. (U.T. is Universal Time, which basically corresponds to British time, five hours later than Eastern Standard Time and eight hours later than Pacific Standard Time.) Thus, the transit starts at 2:13 P.M. E.S.T. on the East Coast and 11:13 E.S.T. on the West Coast. The Sun will set with Mercury still transiting.

Most transits of Mercury have been observed only with ground-based telescopes. But telescopes in orbit to observe the Sun have observed the transits of Mercury of 1993 and 1999. The 1993 transit was observed by the Japanese x-ray telescope known as Yohkoh (Sunbeam), using an American-built camera aboard. Indeed, Mercury crossed in front of the solar corona, which was continuously visible from the spacecraft but which is not generally visible from Earth; this transit does not appear in ordinary tables of transits of Mercury.

The 1999 transit was observed with the Transition Region and Coronal Explorer, known as TRACE, a NASA spacecraft built by the groups of Alan Title at Lockheed Martin's space sciences laboratory and Leon Golub at the Harvard-Smithsonian Center for Astrophysics. I spent a sabbatical year in which I worked with Golub, and my appreciation of the problem of the black-drop effect was enhanced through a historical paper delivered at an American Astronomical Society meeting by Brad Schaefer, then at Yale and now at the University of Texas at Austin. He pointed out that most of the sources he examined, both books and scientific articles, mistakenly claimed that the black-drop effect was caused by Venus's atmosphere.

I brought the suggestion of examining the TRACE transit of Mercury to Glenn Schneider, a scientist at the Lunar and Planetary Laboratory at the University of Arizona who works intensively with one of the cameras on the Hubble Space Telescope. Golub, Schneider, and I presented a paper with our results at a meeting of the Division of Planetary Sciences of the American Astronomical Society, held in New Orleans.

Perhaps a surprise was that even the transit of Mercury observed from TRACE still showed a black-drop effect. After all, Mercury has no atmosphere (or, at least, a completely negligible one). Furthermore, TRACE was observing from outside Earth's atmosphere, so we knew that Earth's atmosphere was not contributing to this particular black-drop effect.

By modeling the effects in a computer, we reported that the black-drop effect observed for Mercury came from two sources:

◆ The instrument involved a small telescope, which had its own fundamental limitation to the clarity of its view. The resulting blurring contributed.

◆ The Sun shades off in darkness near its edge.

The blurring from the instrument compounded with the darkening near the Sun's edge provided all the observed black drop, even without any atmosphere on Mercury.

What is this darkening near the Sun's edge? It is known as limb darkening, since the edges of the Sun and stars are known as their limbs. We see into the Sun until its gas gets too murky for us to see farther. When we look at the center of the Sun's disk, we see in as far as a certain level that we call the Sun's surface. When we look near the limb, though, we are looking diagonally. The murkiness adds up so that we can't see any farther at a point that is above the Sun's surface. Then we see a brightness that corresponds to the temperature of the gas at that point and regions near it. From the fact that the points near the edge look darker than the points at the center of the Sun's disk, we deduce that they are cooler. We know this since cooler gas is not as bright as hotter gas.

Thus, we have shown the black-drop effect for a transit of Mercury to arise from the instrumental blurring plus the Sun's limb darkening. A transit of Venus must have the same contributing factors. For observations of transits of Mercury or Venus from Earth, the blurring from Earth's atmosphere can contribute as well. But Venus's atmosphere is not sufficiently thick in size to contribute substantially.

Our Time Has Come

Our generation on Earth will be fortunate to see transits of Venus, even though nobody now alive has ever seen one. The oldest people on Earth are about 115 years old, according to *Guinness World Records,* and it will have been a 122-year gap.

The Transit of 2004

On June 8, 2004, the beginning of the transit will be visible from a large part of the world, from the Middle East eastward through all of Asia. The transit will take about six hours. The Sun will rise with the transit already begun for the eastern part of the United States and Canada, as well as the eastern part of South America and the western part of Africa. The Sun will set with the transit going on for people on the west coast of China and Japan, and people in the South Pacific, including Indonesia and Australia.

In between, including essentially all of Europe and Asia, the whole six hours of transit will be visible. Weather predictions based on past statistics of clear sky show that sites in Egypt may be especially favored, though the event will be high in the sky throughout the entire Middle East across as far as India.

The transit of Venus of 2004.

(F. Espenak, NASA's GSFC)

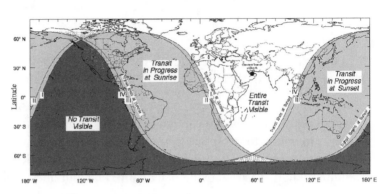

The Transit of 2012

The circumstances on Earth for 2012 will be almost the opposite of those for the 2004 event. The transit will have begun as the sun rises in almost all of Europe and in Asia as far as mid-China and Indonesia. From mid-Australia and mid-China and Russia, through Japan, and past Alaska, the entire transit will be visible. The Sun will set while the transit is going on throughout the rest of the United States, including Hawaii and the whole mainland, also including Mexico and Central America.

The transit of Venus of 2012.

(F. Espenak, NASA's GSFC)

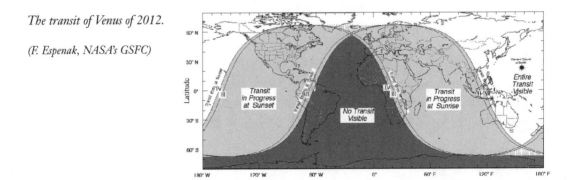

The transits of Venus won't be as spectacular to the eye as total eclipses of the Sun. Nevertheless, as intellectually and historically interesting events, they will rank very high.

Solar Scribblings

As we go out into the solar system with robotic and crewed spacecraft, we will have chances to see other objects transit the Sun. From Mars, one of its two moons, Phobos and Deimos, will transit every week or so. From Jupiter, five of its satellites will hide the Sun entirely. Remember that Jupiter is five times farther from the Sun than Earth is, so the Sun appears five times smaller than it does from Earth. From Saturn, the Sun appears 10 times smaller than it does from Earth, and four of its satellites entirely hide the Sun; transits of others will be observable. The transit of Earth as seen from Saturn on January 13, 2005, will be difficult to detect even for the Cassini spacecraft that will then be in orbit around it. And Cassini is not allowed to be pointed toward the Sun. The Sun appears so small from Uranus, Neptune, and Pluto that transits by most of their moons would be total eclipses.

The Least You Need to Know

♦ Halley and others calculated how transits of Venus could be used to measure the size of the solar system.

♦ Eighteenth-century expeditions were many, but their success was limited by the black-drop effect.

♦ Modern studies of transits of Mercury have shown that the black-drop effect results from the darkening of the Sun near its edge and from the limited resolution of telescopes.

♦ The transits of 2004 and 2012 will be contemporary chances to repeat historical accomplishments.

Part 4

The Sun from Mountaintops

Astronomers ride free on the ski resort's cable car, going one stop higher. How high and where do we go for the clearest and best view of the Sun through the daytime sky? As Earth turns, cooperating observatories around the globe take turns watching it—just to make sure it doesn't get away with anything. And by following it so closely, astronomers even figure out what the Sun is hiding, both inside and around its back.

Chapter 17

High Above the Clouds

In This Chapter

- ◆ Solar versus nighttime observatories
- ◆ Seeing versus transparency
- ◆ Putting solar telescopes on towers or out in lakes
- ◆ Altitude and infrared

Most people's mental pictures of observatories involve giant telescopes peering up at the night sky. But the Sun is up only in the daytime, a profound observation that changes everything. Scientists observing the Sun have to peer through an atmosphere roiling with the heat from the very object they are trying to study. Observing the Sun requires more differences from traditional astronomy than merely getting up in the morning instead of staying up all night.

The Sun Shines in the Daytime

On a dark night, if you are far from lampposts and other city or town lights, you can see perhaps 3,000 stars. Each of those stars is a mere point of light to us on Earth. Even using the largest telescopes in a straightforward manner does not enlarge the images of stars; the images remain points. Only a special technique now being intensively developed, involving linking separate but adjacent telescopes, has been able to detect structure on the surfaces of any stars or to measure the sizes of stars. And the Hubble Space Telescope has been able to detect only giant blobs of structure on the star Betelgeuse, a supergiant star in the constellation Orion that is relatively close to us.

A bright area on Betelgeuse, a supergiant star in the constellation Orion, imaged with the Hubble Space Telescope. This amount of detail is the most ever seen on the surface of a star other than the Sun.

(Andrea Dupree [Harvard-Smithsonian CfA], Ronald Gilliland [STScI], NASA, and ESA)

Telescopes that study stars, therefore, are used as light buckets, collecting as much light as possible and funneling it to a small area on a detector. The light from a star often is directed through a spectrograph, which spreads the light into its component colors. Or, it may be viewed through a series of broadband filters, each of which passes a range of colors. But in all those cases, all the light from each star is treated together. No detail on a star's surface can be analyzed.

Fun Sun Facts

The Sun covers about ½° across the sky. (Scientists often say that it subtends—that is, covers—½° of arc.) Compare this angle with the 360° of a full circle. The width of your thumb at the end of your outstretched arm covers about 2°, so your thumb more than covers the Sun.

In these standard angular measures, each degree is divided into 60 minutes of arc, and each minute of arc is subdivided into 60 seconds of arc. So a degree contains 60 × 60 = 3,600 seconds of arc, and the Sun's diameter covers around 1,800 seconds of arc. Telescopes on Earth can resolve detail about 1 second of arc, so 1,800 pixels, each 1 second of arc across, cover the Sun's diameter. A two-dimensional image of the Sun covers roughly πr^2, the area of a circle, with radius (r) equal to 900 arc seconds, so the Sun's disk covers roughly 2,500,000 arc seconds.

The case for the Sun is very different. Compared with the 360° of a full circle, the Sun covers a half degree in the sky. Scientists and mathematicians talk of parts of a circle as "arc," and divide degrees into 60 minutes and each minute into 60 seconds. This half degree is, in turn, the same as 30 minutes of arc. Unaided, the human eye can see detail about 1 minute of arc across (we often say "1 arc minute," for short). That would be about the size of a moderately large sunspot. To see any smaller angle, we need a telescope.

> **⚠ CAUTION**
>
> ### Sun Safety
> Even though sunspots seen with the naked eye have been reported for hundreds of years at sunset, when the Sun is sufficiently dimmed by Earth's atmosphere, you must look at the Sun only through a special solar filter that brings the sunlight intensity down to a safe level.

In principle, the larger a telescope is, the finer the details it can resolve. Something the size of the pupil in the human eye, which is perhaps 6 mm across, resolves about 1 arc minute, $\frac{1}{30}$ of the Sun's diameter. Something 60 times bigger resolves detail about 60 times smaller—that is, 1 arc second. And 60 times bigger than 6 mm is 360 mm, which is about 15 inches. But at about 1 arc second, the turbulence in Earth's atmosphere keeps you from seeing finer detail. An average-quality observation site for an amateur astronomer might be limited to detail larger than about 3 arc seconds. A high-quality professional site is traditionally said to have a limit of about 1 arc second. The very best sites, such as the Mauna Kea Observatory, at 4,215 meters (13,800 feet) of altitude in Hawaii, has "seeing" (image steadiness) that reaches perhaps 0.4 arc seconds. The Hubble Space Telescope can see about 0.1 arc seconds in the visible part of the spectrum.

Because of the limitations of the Earth's atmosphere, it has not been generally thought important to build very large solar telescopes. After all, the Sun is so bright that we usually get enough light even with a smaller telescope. Only at the present time is there a major project underway to build a solar telescope of a size comparable to that of even the previous generation of large nighttime telescopes, as we discuss in Chapter 19.

The Sun Isn't Up at Night

The air isn't usually as steady during the day as it is at night. After all, energy from the Sun is coming down to Earth through the air, heating it. Columns of rising air have different densities from surrounding air, which makes the light passing through them bend every which way. Therefore, images during the day are ordinarily not as clear as images at night.

The property of unsteadiness in images that is introduced by turbulence in the air is known as *seeing*. Bad seeing introduces a couple of effects. One is called dancing, in which the image moves around the sky a little. If you are taking a photograph, the image will blur because of its motion during the time that the shutter is open. (In some cases, you can take a lot of very short exposures, each of which has finer resolution, and add them together later after aligning them.) Another is straightforward blurring.

Sun Words

Seeing is the quality that describes how steady images are.

When the adaptive optics system was turned on at the Keck 2 telescope, to counteract the effects of seeing by distorting one of the telescope's mirrors, the image size improved by a factor of over 10. This infrared image of a star improved from the blurry half an arc second (left) to the more pinpoint ¹/₂₀ of an arc second (right); both images are shown to the same scale.

(Keck II AO Facility/NASA/LLNL)

Solar Scribblings

Astronomers hope for "good seeing" when they go observing. A polite way to send off a friend to an observing session is "I hope the seeing is good." That means that you are hoping for steady air that will allow good resolution on the images.

Another effect that astronomers worry about is *transparency*. Transparency is how clear the sky is—that is, what fraction of the incoming light gets through the atmosphere. The transparency varies from day to day and also varies with angle in the sky. Near the horizon, you are looking through several times as much air as you are when you look straight up, so the transparency goes down. Astronomers say that there is extra *extinction*.

Solar astronomers hold up their thumbs a lot against the sky, blocking out the Sun to see how clear the sky is. On a clear mountaintop site, the sky can appear steady blue up to the edge of your thumb. From sea level or close to it, you can usually see the sky getting brighter and whiter as you get closer to your thumb (and to the solar disk that is hidden by it). Never at sea level do you have a sky that is as deep blue as it is at a high altitude.

Sun Words

Transparency is how clear the atmosphere is.
Extinction is how much of incoming light is absorbed before it reaches us.

It is possible to have good seeing and bad transparency; conversely, it is possible to have bad seeing and good transparency. For example, an atmospheric inversion can cause smog to settle over a city—for which the Los Angeles area has been famous. That smog lowers the transparency, but the stagnant air can lead to good seeing. So a smoggy day can have good seeing yet bad transparency. On the other hand, a day that is made clear by wind whipping around can have blurred images because of bad seeing, yet it can have good transparency.

Solar Scribblings

Some of the best solar movies of all time were made in an asphalt-paved parking lot in Burbank, California, a far cry from the scenic mountains or mountain lakes of other solar observatories. The site was chosen because the Lockheed Corporation, an aerospace company, supported a group of scientists studying the Sun, and it was convenient for them to observe right out their back door. They called their location the Lockheed Solar Observatory. Because of the smog, the site often had excellent steady seeing over periods of hours, even though the transparency was lousy.

Higher Telescopes

One way of eliminating the turbulence that comes from solar heating of the ground is to put the entrance aperture of your telescope at the top of a tower and build the telescope as an integral part of the tower. Two solar towers at the Mt. Wilson Observatory, overlooking Los Angeles, have been there for almost 100 years. The original tower is 60 feet (18 meters) high. George Ellery Hale discovered the magnetic field of sunspots with this tower telescope in 1908. The other tower is even higher, 150 feet (45 meters), to provide higher dispersion in the spectra it takes. Since the Sun is up in the daytime, the bright nighttime lights of today's Los Angeles aren't relevant to solar observation.

Scientists using the towers continue to make daily magnetic-field images of the Sun as well as other images. Some of the measurements are used for helioseismology. In each case, the tower is only part of the telescope; the light is reflected downward from mirrors at the top of the tower through a lens and into a pit deep below the mountaintop. It is then bounced back up to the observing room, which is at mountaintop level.

The 60-foot and 150-foot solar tower at the Mt. Wilson Observatory, now operated by UCLA.

(UCLA)

Not all mountaintops are good for observatories. To have a good site, you must have not only altitude, but also a steady flow of air. The Hawaiian shield volcanoes, such as Haleakala on the island of Maui and Mauna Loa on the island of Hawaii, have such gentle slopes that the air flows in a calm, laminar fashion across and above them. Thus, solar observatories are located at each of those places. But random mountains like Pike's Peak may have turbulent air above them, and we don't put solar observatories there.

Cutting Down on Turbulence

Another way to get a smooth flow of air above your observatory is to put it in the middle of a lake. The best way to do so is to put it on a small island, though you can make an artificial island, as was done in the middle of Big Bear Lake in the San Bernardino Mountains a couple of hours west of Los Angeles in California. Because of its location in the lake, the seeing is steady for long periods of time, and the Big Bear Solar Observatory is famous for movies of changing solar phenomena.

Solar Scribblings

The story has it that astronomers realized that lakes might be useful sites for solar observatories from an observation made at the Sacramento Peak Observatory in the mountains above Alamogordo, New Mexico. Someone noticed that the seeing seemed to improve at one of the telescopes when they were watering the lawn. It was but a small step from watering the lawn, with evaporating water diminishing the turbulence of the air flowing above it, to seeking out a full lake.

Turbulent air makes bad seeing not only above the telescope, but also in the telescope itself. The hot beam of solar light causes rising and falling air currents within the telescope's tube. A first round of rejection is a filter that cuts out the infrared part of the solar light. Such a filter has to be placed at the front of the telescope tube. Second, fans within the tube might distribute the air, to keep it from turbulence. The most drastic solution is to evacuate the telescope tube, making a vacuum inside. Devoid of air, the inside of the tube cannot contribute to turbulence.

Let It All Come Through

Our eyes are most sensitive at the colors of light that the Sun emits most strongly—especially in the orange, yellow, and green. But other kinds of light exist that we don't see. In particular, beyond the red end of the spectrum, we have infrared.

Solar Scribblings

Even though William Herschel had discovered the planet Uranus in 1781, he did not rest on his laurels. About 200 years ago, Herschel was experimenting with sunlight, spreading it into a spectrum. He put a thermometer into the different colors to check their temperatures. He was surprised to find that his thermometer showed an excess temperature even when he held it beyond the red, compared with off to the side. He had discovered infrared, named from the Latin *infra-*, meaning "beyond."

Many things in solar astronomy are studied better in the infrared than they are in the visible. The magnetic field of sunspots, for example, shows up particularly well in infrared spectra. But infrared observations have lagged behind optical observations in all kinds of astronomy for both atmospheric and instrumental reasons.

The instrumental reason is that particles—photons—of infrared light each contain less energy than individual photons of visible light. The infrared photons, in particular, do not have enough energy to affect ordinary film. Though some infrared films exist that barely detect the part of the infrared immediately adjacent to the red, the infrared extends far beyond that. Observing infrared well had to await the invention of sensitive electronic detectors for that part of the spectrum.

Solar Scribblings

The first infrared detectors had only one detecting element, but at the dawn of the twenty-first century, infrared detectors with hundreds of pixels existed. This resolution is still far short of optical detectors, which now contain millions of pixels. Still, the infrared space observatory that NASA planned to launch in 2003 to be equivalent in scale to the Hubble Space Telescope and the Chandra X-Ray Observatory should be able to produce reasonably detailed infrared images. Another infrared camera was rejuvenated in 2002 aboard the Hubble Space Telescope. But neither of these infrared devices can look at the Sun without being blown out. Only special solar telescopes and spacecraft are ever purposely pointed at the Sun.

The atmospheric reason for limitations in infrared observing stem from the fact that water vapor absorbs infrared. Since there is water vapor in the Earth's atmosphere, little of the infrared comes through to the Earth's surface. Though the atmosphere is opaque to most of the infrared, some bands of infrared color come through. We say that they come through "windows of transparency," though the windows are merely in the spectral coverage.

The higher and drier the site is, the better infrared radiation can be observed. So observatories at high sites, like the Mauna Kea Observatory at 4,215 meters (13,800 feet) of altitude, have telescopes optimized to observe the infrared. The Mauna Loa Observatory, on a Hawaiian mountain facing Mauna Kea, is similarly high. But only in recent years have infrared detectors developed enough to make it worthwhile to devote a lot of effort to infrared solar observations.

Infrared can be observed well from high-flying airplanes. NASA had its Kuiper Airborne Observatory aloft through the late 1990s, but it was retired to devote the budget to next-generation aircraft: Stratospheric Observatory for Infrared Astronomy (SOFIA), a joint project of NASA and the German Space Agency. SOFIA carries a 2.5-meter (100-inch) telescope, larger than any previous telescope aloft or in space, and is to start its work in 2005. Though the Kuiper Airborne Observatory carried out most of its observations pointed at objects other than the Sun, it did make some notable solar observations. For example, it was used to observe infrared radiation near the limb of the Sun during a solar eclipse.

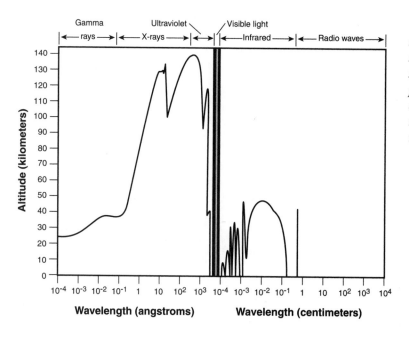

Windows of transparency, showing which parts of the spectrum reach the Earth's surface. The whole visible part of the spectrum, but only narrow bands in the infrared, are in windows of transparency.

The Least You Need to Know

- ◆ Solar observatories need different conditions than nighttime ones.
- ◆ Good seeing and good transparency are not necessarily linked.
- ◆ Solar telescopes can be on towers, lakes, or high mountains.
- ◆ New instrumental capabilities are improving infrared solar observations.
- ◆ High altitude is needed for the best infrared observations.
- ◆ An instrumented airplane will fly high in order to observe the infrared.

Sunspot, New Mexico, and the House of the Sun

In This Chapter

- ◆ The National Solar Observatory's telescopes in New Mexico and Arizona
- ◆ Telescopes on Maui's Haleakala crater
- ◆ Observatory of the Lake
- ◆ Solace from SOLIS

Astronomy is an international profession, with people all over the world studying the Sun. Some of the major telescopes are in the United States, and we discuss them in this chapter. In the following chapter, we discuss some of the international sites, as well as major future ground-based projects.

Southwest Slant

The telescopes of the United States National Solar Observatory are in Sunspot, New Mexico, and on Kitt Peak, near Tucson, Arizona.

The Largest in the World

The largest solar telescope in the world—that is, the one with the largest focusing mirror—is on Kitt Peak, alongside the nighttime telescopes of the United States National Optical Astronomy Observatory. It is the McMath-Pierce Telescope of the National Solar Observatory (NSO).

The telescope is so large that it doesn't move around the sky, pointing at the Sun as it traverses the heavens each day. Instead, it has a pair of large, flat mirrors at its upper end. They reflect the sunlight into the focusing part of the telescope.

In the McMath-Pierce Solar Facility, on Kitt Peak near Tucson in Arizona, a flat mirror 2.1 m (7 feet) across reflects sunlight down the slanted tube and into an underground chamber. The observing room is at ground level.

(National Solar Observatory/AURA/NSF)

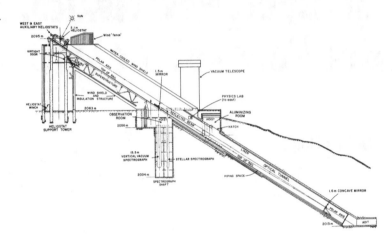

Hold Still

In the mid-nineteenth century, French astronomer Jean Bernard Léon Foucault thought of a way of keeping stars steady in a telescope without moving the telescope itself. His device was called a siderostat, from the Latin words for "star" and "steady."

Fun Sun Facts
The second-largest siderostat telescope, and the third-largest that has ever existed, is at the Buhl Planetarium in Pittsburgh, Pennsylvania. Its lens is 25 cm (10 inches) across. A 1.5-m (5-foot) siderostat was built by the Observatory of Paris and exhibited in Paris in 1900 with the largest refracting telescope ever built. It was disassembled at the end of the Paris Exhibition and hasn't been used since.

Out of Air

The McMath-Pierce Solar Facility may be one of the world's largest solar telescopes, but it isn't at the best site for solar observing. One reason is that at 2,100 m (6,875 feet) of altitude, Kitt Peak is not a very high mountain, as astronomical observatories go, which makes it not a good site for a coronagraph. The sky gets purer blue, among other places, at Sacramento Peak in the Sacramento Mountains high above Alamogordo, New Mexico. The site is at an altitude of 2,800 m (9,200 feet) in a beautiful pine forest.

The observatory was set up in the 1940s by Donald H. Menzel of the Harvard College Observatory. Menzel was a noted theoretician, famous especially for his studies of the solar chromosphere. He also built several of the best solar observatories. The optical designer Richard B. Dunn has long been resident at Sacramento Peak, and he designed a most unusual and successful telescope. It was originally called simply the Vacuum Tower Telescope but has since been named the Richard B. Dunn Solar Telescope.

Solar Scribblings

The wavelength in angstrom units (10 millionths of a meter) of the red spectral line of hydrogen, often used in observing the Sun, is 6,563. As a joke, the number of the New Mexico highway that leads to the Sacramento Peak Observatory is, most unusually, the same 6,563.

The Richard B. Dunn Solar Telescope, whose insides are a vacuum to eliminate the solar heating that blurs images.

(National Solar Observatory/AURA/NSF)

At the top of the Dunn telescope, a pair of flat mirrors directs the solar beam downward. A quartz window seals the top of the vacuum. A negative consequence of using the flat mirrors is that the image rotates. This rotation, once per day, would blur the highest-resolution images unless it was compensated for. Dunn compensated brilliantly—and expensively—by suspending the whole working part of the telescope from a mercury bearing near the top of the tower. The light beam goes deep into the basement and is focused back up at the ground level, where the observing floor with its cameras is part of the rotating part. The telescope's lens is 76 cm (30 inches) across.

Maui's House

Haleakala, the Hawaiian name for "House of the Sun," is a volcanic crater 3,000 m (10,000 feet) high on the island of Maui. (In Hawaiian, *hale* is "house" and *La* is the Sun god.) Maui is a few hundred miles down the chain of Hawaiian islands from Oahu, where Honolulu, the capital of the state, is located.

 Solar Scribblings

In Hawaiian legend, the god Maui lifted up the sky to the heavens. Maui went up to the biggest volcanic crater on the island now named after him, to the house of La, the Sun god. Maui caught La and got him to agree to travel more slowly over the island during half of each year.

Before they founded an observatory on the "Big Island" of Hawaii, University of Hawaii scientists started a solar observatory on Maui. Now called the Mees Solar Observatory, this site boasts especially clear air overhead, enabling researchers to operate coronagraphs to study aspects of the solar corona even without total eclipses.

The C. E. K. Mees Solar Observatory of the University of Hawaii is at 3,054 m (about 10,000 feet) of altitude. It is so high that you should be careful not to run when you first go up, lest you become faint or dizzy in the thin air.

The observatory runs some monitoring programs, including using a dual coronagraph to study not only the inner corona, but also the chromosphere and prominences. It does so by observing with coronagraphs. By *coronagraph*, we mean a system that blocks out the ordinary Sun in the middle of the image, even though, in this case, one of the instruments is used to observe the chromosphere rather than the corona.

The University of Hawaii scientists also have a special telescope to measure the magnetic field of the Sun in detail. And they measure the polarization of the solar phenomena such as sunspots.

One of their telescopes monitors the solar surface in white light. Another is an imaging spectrograph that records the spectrum of all points in a square region about 200,000 km (about 125,000 miles) on a side. It does so repeatedly, to look for changes.

In the next section, we will mention a solar telescope on a different Hawaiian island, the so-called Big Island.

Corona Without an Eclipse

It wasn't until the mid-nineteenth century that astronomers were convinced that the halo seen around the Sun at eclipses was really part of the Sun instead of being an effect in the Earth's atmosphere. You learned in Part III how scientists in 1868 (in one case spurred by eclipse observations) succeeded in viewing the chromosphere without an eclipse. But the corona is about another thousand times fainter.

Attempts to view or photograph the corona without an eclipse did not succeed for many decades. Then, in the 1930s, French optical scientist and astronomer Bernhard Lyot (pronounced *lee-oh*) brilliantly succeeded through elegant and careful work. His coronagraphs used only lenses; the surfaces of mirrors had too many pinholes and other irregularities that led to excessive scattered light. It didn't take too much light scattering around to drown out the faint light from the corona.

One of the keys to Lyot's success was his careful attention to where the light went when it was scattered off the edges of the lenses or off the lenses themselves. By focusing the solar image onto an intermediate surface and blocking the extraneous light, Lyot went a long way to taming the scattered light in the system.

Lyot also paid careful attention to using the best optics with the smoothest possible surfaces, and not having dust or other material on the lenses that would scatter light. Only with his designs and fabrication did he succeed where others failed in building a device that allowed observation of the corona without an eclipse.

Solar Scribblings

To keep the lenses in coronagraphs smoothly coated, the Sacramento Peak Observatory director Jack Evans recommended "nose oil." Before I actually went to the store to ask for some, perhaps confusing it with "3-in-1 oil," I discovered that he meant the actual oil from the side of his nose. He took a soft cloth, touched it to his nose, and transferred the small bit of oil to one side of the coronagraph lens. He then gently rubbed in larger and larger circles, spreading the tiny bit of oil out over the whole lens. Much of astronomy and of other science depends on lore and traditional techniques like this. I don't see a computer program simulating nose oil!

The Solar Scoop

The intensity of the corona is often measured in units of millionths of the intensity of the middle of the solar disk. The inner part of a bright corona is about 50 millionths. So the sky has to be fainter than 50 millionths to allow for viewing the corona. That condition is never met at low altitudes but is common at Haleakala.

Coronagraph Tricks

Coronagraphs are often used with filters that pass only the tiny part of the spectrum in which a coronal emission line, such as the coronal red or green line, is present. Then the corona is perhaps 10 times brighter than in other parts of the spectrum, relative to the sky. Still, only the inner part of the corona is visible. Coronagraphs on Earth do not see the corona out as far as one can see it at eclipses.

Solar Scribblings

The word *coronagraph* is now often used for any device that blocks out a bright central object. For example, the Advanced Camera for Surveys that astronauts installed in 2002 on the Hubble Space Telescope has a coronagraph mode to allow faint objects to be seen near bright ones. It allows dust disks that might contain planets in formation to be seen around other stars. It has also looked into the galaxies that are hosts to tremendous eruptions known as quasars.

Fun Sun Facts

NASA is considering a mission called Eclipse to use a kind of coronagraph on a 1.8-m (6-foot) telescope in orbit in order to try to image planets around other stars. It would be a precursor to NASA's Terrestrial Planet Finder mission.

The inner part of the corona that we see at eclipses is caused by solar photospheric light scattered by electrons in the corona. The scattering process polarizes the light highly. By examining the polarization, coronagraphs can be used to study the corona farther out than otherwise. Such a coronagraph is present at the Mauna Loa Observatory in Hawaii. It maps the corona every clear day at distances above the solar limb from 1.08 to 2.85 solar radii. It takes three minutes to construct each image.

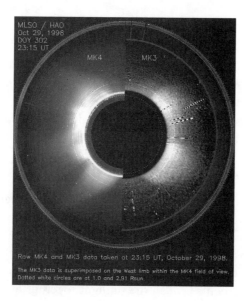

MLSO / HAO
Oct 29, 1998
DOY 302
23:15 UT

MK4   MK3

Raw MK4 and MK3 data taken at 23:15 UT, October 29, 1998.
The MK3 data is superimposed on the West limb within the MK4 field of view.
Dotted white circles are at 1.0 and 2.91 Rsun.

Comparison of the coronal images from the Mark 3 and 1999's Mark 4 improvement in the coronagraph at the Mauna Loa Observatory in Hawaii. The coronagraph images white light.

(Mauna Loa Observatory, High Altitude Observatory, NCAR)

California Dreaming

After an hour's drive in the flat ground east of Los Angeles, you start to drive up into the mountains. As you go higher, you reach a pine forest. There, on a resort lake, you see a strange white building and dome out from the shore. This is the Big Bear Solar Observatory.

The site is 2,040 m (6,700 feet) high, not high enough to be a coronagraph site, but high enough to get clean air for high-quality images. Using modern computer technology, observatory scientists not only take their images digitally, but also make them immediately available on the World Wide Web. You can see the Sun minute by minute through web pages at www.bbso.njit.edu/.

Big Bear is about to undergo a major upgrade. The New Jersey Institute of Technology and the University of Hawaii are together replacing the existing largest telescope with a much larger one, 1.6 m (65 inches) in diameter. It will use the latest in adaptive optics, flexing a mirror to counteract the blurring effect of the Earth's atmosphere. They expect to obtain resolution of 0.2 arc seconds in the infrared and even better in the visible when the telescope goes into use in 2005. Among the new projects will be high-resolution observations of solar flares at a rapid retition rate.

Synoptic SOLIS

Solar scientists like to make synoptic observations, which simply means making observations over a long period of time. Such long sets of data allow scientists to study long-period or long-term variations on the Sun. The United States National Solar Observatory has built such a system to study how the sun changes over the sunspot cycle.

SOLIS stands for Synoptic Optical Long-term Investigations of the Sun. In addition to studying how the Sun's magnetic field and features change over the sunspot cycle, it supplies information about the relation of flares and other solar activity to the Earth. One 14-cm (6-inch) telescope of SOLIS measures the details of structure on the full solar disk. Another component telescope measures the magnetic field in detail, showing its direction in three dimensions in addition to its strength. This telescope, the largest, is 50 cm (20 inches) across and makes these observations with a resolution of 2 arc sec. Other instruments take all the sunlight together but measure its intensity at various wavelengths very accurately, complementing space observations of the change of the overall solar radiation; the goal is to show how much the amount of energy we receive from the Sun changes over time, both over short time scales and over the 11-year sunspot cycle. The light for these last observations is obtained with a tiny telescope only 8mm (⅓ inch) across. All three telescopes are on a single mount.

SOLIS makes a wide range of observations daily through filters that pass light from different elements and measure different aspects of the solar atmosphere.

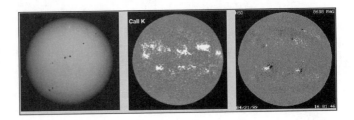

SOLIS observes, among other things (left to right), white light, the chromosphere (by observing the so-called K line of ionized calcium), and the magnetic field, with opposite polarities shown in white and black, respectively.

(National Solar Observatory/AURA/NSF)

After it is calibrated, SOLIS will replace the Vacuum Tower Telescope on Kitt Peak and some instruments on Sacramento Peak. But it is hoped that duplicate SOLIS facilities will be built at other longitudes around the world, to provide continuous high-quality observations of the Sun over 24 hours. SOLIS itself is planned for operation over at least a full double sunspot cycle of 22 years.

SOLIS was the highest priority of a 1990s report on the needs of the solar-astronomy scientific community. It will show its capabilities in a variety of ways that correspond to the new scientific methods of doing things. It will be operated remotely, and its data will be quickly available over the Internet, with no restrictions on its use. So you will be able to use the data from SOLIS to study the Sun as quickly as any professional solar astronomer could.

The Least You Need to Know

- Sunspot, New Mexico, and Kitt Peak, Arizona, have our national solar telescopes.

- Telescopes on an island in Big Bear Lake, California, make solar movies.

- A high crater in Hawaii hosts an excellent site for studies of the corona.

- Special and careful techniques allow observation of the inner corona without an eclipse.

- A new telescope is being prepared for long-term solar studies.

Chapter 19

Canaries and the Big Dog

In This Chapter

- ◆ The newest, best solar telescope in the Canary Islands
- ◆ Near the Lake Palace
- ◆ Are the Alps in Japan?
- ◆ A wandering telescope I

Just as the Sun seems to go around Earth once a day, at least from our point of view on the ground, solar astronomy flourishes at all longitudes. Efforts from a wide variety of countries—including Sweden, India, and Japan—give at least a brief picture of the range of interesting things to tackle in order to understand our Sun.

Gone to the Dogs

The Canary Islands are off the west coast of Africa, about where that continent bulges farthest to the west. They have been part of Spain for hundreds of years. Early European visitors, including Pliny the Elder, about 2,000 years ago, mentioned the many big dogs found there. Since the Latin for "dogs" is *canes*, the location was named the Islas Canarias, which has been translated as the Canary Islands.

Two of the Canary Islands have major observatories atop mountains. The headquarters of the Canaries Institute for Astronomy and Astrophysics (Instituto de Astrofísica de Canarias) is on Tenerife, as is one of the observatories. The other observatory, the Roque de los Muchachos Observatory, contains the biggest nighttime telescopes and is on La Palma. A Spanish telescope equal in size to the 10-m (400-inch) Keck telescopes, the largest in the world, is opening soon on La Palma.

The newest and perhaps the best of the world's solar telescopes opened on La Palma in 2002. It is run by the Institute for Solar Physics of the Royal Swedish Academy of Sciences in Stockholm, Sweden. This Swedish Solar Telescope has a focusing lens 1 m (40 inches) across. That makes it the second-largest optical solar telescope in the world. The lens also acts as the top element that closes off a vacuum. The light path inside the telescope is evacuated.

Fun Sun Facts

The 1-m (40-inch) lens of the Swedish Solar Telescope is a single piece of glass and, thus, focuses different colors at different distances. Therefore, most observations are made using only one wavelength at a time, chosen with a filter or with a spectrograph. A more complicated set of lenses is also available to allow several wavelengths to be combined at the same focus.

The Swedish 1-m (40-inch) Solar Telescope at the Roque de los Muchachos Observatory on La Palma, Canary Islands, Spain.

(Royal Swedish Academy of Sciences)

Fun Sun Facts

Vacuum telescopes are found all over the world, including on Kitt Peak in Arizona, on Sacramento Peak in New Mexico, at Baikal in Russia, at Udaipur in India, and in Huairou and Kunming in China. Huairou is the solar observation station of the Beijing Astronomical Observatory of the Chinese Academy of Sciences (http://sun.bao.ac.cn/) and was opened in 1997. It is on a small island in a reservoir about 60 km (40 miles) from downtown Beijing.

The telescope was designed from scratch to give the finest possible solar images, approaching one tenth of an arc second. Göran Scharmer, already well known for his high-resolution imaging with the telescope's half-size predecessor, designed the telescope and its imaging system. To limit turbulence in the telescope, it is evacuated. To reach the desired resolution, about 10 times better than the traditional limit, it incorporates adaptive optics. This latter method uses a mirror that changes in shape as often as 1,000 times per second. The shapes are calculated in real time by a computer that interprets the blurring of the image to show the shape that is needed to compensate for turbulence in the air outside the telescope.

The telescope has already shown its success in making high-resolution images of sunspots.

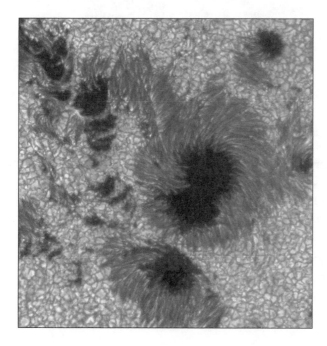

High-resolution images of sunspots made with the Swedish Solar Telescope. The dark umbra is surrounded by fibrils of the penumbra. The salt-and-pepper effect known as granulation, part of the photosphere, surrounds the sunspot.

(Royal Swedish Academy of Sciences)

Indian Idyl

In India's state of Rajasthan is the city of Udaipur, about 800 km (500 miles) south-west of New Delhi and northwest of Mumbai (formerly Bombay). Its lake is the site of the Lake Palace Hotel, known to many moviegoers as the site of one of the James Bond spectaculars, *Octopussy*. Guests take a boat from the shore out to the island where the hotel is located.

The Udaipur Solar Observatory is in the middle of a lake.

(Jay M. Pasachoff)

But boats go elsewhere in the lake as well. When Arvind Bhatnagar returned to India from years of supervising the Big Bear Solar Observatory, he vowed to build another island solar observatory to provide high-quality seeing. And he succeeded. The observatory has since become part of the Physical Research Laboratory of the Department of Space in India, whose headquarters are in Ahmedabad.

Like so many solar observatories, the Udaipur Solar Observatory has several telescopes on a single mount. The largest telescope has a 25-cm (10-inch) lens and is used for high-resolution observations in hydrogen light. Another telescope has a 15-cm (6-inch) lens and is used for hydrogen-light full-disk observations.

India has several astronomical observatories. A coronagraph site is being planned at the very high altitude of 4,500 m (1,475 feet) in the Himalayas. Indian scientists have also been active in observing eclipses.

The Udaipur Solar Observatory is one of the sites of the GONG network, which we describe in the following chapter.

The Other Alps

Thus far, we have described mainly solar observatories that make images of the Sun in visible light. But the Sun emits radiation across the entire spectrum from gamma rays through x-rays to ultraviolet, to visible, to infrared, to radio waves.

The wavelengths of radio waves are over 1,000 times longer than the wavelengths of visible light. Since the fineness of the resolution worsens as the wavelength increases, images with optical-size telescopes of radio wavelengths would be truly substandard. But it is possible to build solar radio telescopes spread over acres. These radio telescopes are composed of many separate small antennas, and the signals are combined electronically.

The current best of these radioheliographs is at the Nobeyama Radio Observatory in the Japanese Alps, a five-hour train ride west of Tokyo. Nobeyama is in the guidebooks mainly because it is the railway station of the highest altitude in Japan (1,300 m, or about 7,000 feet). The observatory boasts several large telescopes, including one 45 m (150 feet) across that is the largest radio telescope in the world accurate enough to be used for observations of short-wavelength radio waves, those with wavelengths of only a few millimeters.

Quite noticeable to both astronomers and visitors is a large circle of hundreds of small white dishes. These comprise the radioheliograph. The observations of all these telescopes are combined in a control room nearby. When combined, images of the Sun exist that show activity in the solar corona. The observations are at a frequency of 17 gigahertz (GHz), equivalent to 1.8 cm (0.7 inches) in wavelength. Flares and eruptions show clearly.

Also on the site is a set of a handful of small solar telescopes. Each is sized so that the beam of radiation that it sees comes from a disk the size of the Sun. For those of double the wavelength of an adjacent radio telescope, the diameter is halved, to keep the field of view the same. Decades ago, I worked at such a system at the Sagamore Hill Radio Observatory in Hamilton, Massachusetts, run by the U.S. Air Force Cambridge Research Laboratories.

Some of the telescopes of the Nobeyama Radio Observatory's radioheliograph, with the 45-m (150-foot) millimeter-wave radio telescope in the background.

(Nobeyama Radio Observatory)

Where, Oh Where, Will My Telescope Go?

For too long, astronomers have felt that the Sun is so bright that solar astronomers didn't need a big telescope. But solar astronomers have come to spread out the Sun's light so much in wavelength, or to divide it up into intervals so small in time, that the amount of light in their telescopes limits their observations. They would like a big telescope, like everyone else.

Finally, there is general agreement that solar science has advanced so far that a big telescope is needed. An unprecedentedly large consortium of 22 institutions has banded together to plan for the Advanced Technology Solar Telescope (ATST). It is to be in the 4-m (160-inch) class, a size that matches the nighttime telescopes at the National Optical Astronomy Observatory and the Cerro Tololo Inter-American Observatory, which used to be the largest until they were superseded by telescopes of the 8 m (320-inch) class telescopes.

But it isn't enough just to take over a 4-m (160-inch) telescope for solar purposes. The telescope must be specially designed and optimized for the needs of the solar community. For one thing, it must have adaptive optics designed from the beginning. Note that the area of a 4-m mirror is 16 times (4^2) the area of the 1-m (40-inch) Swedish Solar Telescope, so the ATST would represent a giant leap in abilities. The mirror must have qualities of especially low scattering of the light that hits it. The whole system must function well in the infrared, which involves among other things keeping room-temperature parts of the telescope out of the field of view. And today's generation of cameras and other instruments must have room at the telescopes' foci.

All the major American solar players are participating, with the National Solar Observatory as the principal investigator of the project; the Big Bear Solar Observatory's parent at the New Jersey Institute of Technology, the High Altitude Observatory (based in Boulder, Colorado), and the relevant astronomical parts of the University of Hawaii and the University of Chicago as co-principal investigators. The major support is being requested from the National Science Foundation's divisions of Astronomical Sciences and Atmospheric Sciences. International partners are currently being sought, and a proposal for participation from a consortium of European solar institutions is expected.

An artist's conception of the 4-m (160-inch) Advanced Technology Solar Telescope.

(NSO)

The ATST's consortium has announced that its science program is dedicated to the following questions:

- ◆ What basic mechanisms are responsible for solar variability that ultimately affect human technology, humans in space, and terrestrial climate?

- ◆ How are solar and stellar magnetic fields generated, and how are they destroyed?

- ◆ What role do magnetic fields play in the organization of plasma structures and the impulsive releases of energy seen on the Sun and throughout the universe?

The ATST will be able to observe and follow fine details in the magnetic field structure on the Sun over a wide range of time scales. It will be able to measure three-dimensional polarization and make spectra of very high spectral and spatial resolution. It will take advantage of the experience gained in the current generation of three-dimensional polarization and magnetic field telescopes at such sites as Haleakala, Kitt Peak, and Sacramento Peak. It will also be able to accommodate the advances in adaptive optics, to allow features to be followed on a spatial scale better than one-tenth of an arc second, which is only 70 km (44 miles) at the distance of the Sun from Earth. Numerical calculations show that there should be features this small, and we want to be able to observe them. Some of them are the tubes of magnetic flux that scientists hope to detect coming up through the photosphere. These small columns are key to understanding the formation of sunspots, flares, and other solar phenomena. This high resolution may also allow us to determine which features on the Sun cause the corona to be heated to its temperature of millions of degrees.

Many things are as yet undetermined in planning for the ATST. For example, the best idea so far is to have the tube open at the sides, to allow the flow of air through it. The telescope is too big to consider a vacuum or a helium-filled arrangement. The current idea is that the mirror will reflect the light at an angle to the incoming radiation, to avoid having secondary mirrors in the field of view. The off-axis arrangement would allow for a cleaner image profile and easier accommodation of the very large heat loads from the bright beam of incoming sunlight. The telescope is planned to be sensitive from the ultraviolet just past the visible through the entire visible range and up to about 3 micrometers in the infrared, about four times the longest wavelength (red) of the visible spectrum.

An important part of the planning is to have a very low level of scattered light. In particular, sunspots are dark features on the Sun, and current observations have light scattered from the adjacent, brighter photosphere. In the ATST, the amount of such scattering will be limited to 10 percent of the sunspot's umbral intensity.

The aperture of 4 m (160 inches) is needed not only to collect a lot of light, but also to provide imaging with a spatial resolution of 0.1 arc second at the longer-wavelength end of the infrared range. Also, the large aperture should allow measurements of the three-dimensional magnetic field to be made quickly enough, compared with the rate at which the small magnetic-field elements change.

Though it may be hard to believe, the site for this major telescope has not yet been selected. In some sense, like the Wandering Minstrel in Gilbert and Sullivan's operetta *The Mikado*, the telescope's prospective site is wandering. Site testing is going on all over the globe to provide the best possible location, including excellent seeing and the most hours possible of clear skies. Merely having an existing solar telescope present is not the determining criterion. A half-dozen sites have instruments monitoring conditions over a lengthy period of time: Big Bear Solar Observatory, in California; Observatorio del Roque de los Muchachos, in La Palma, Canary Islands, Spain; Mees Solar Observatory, in Haleakala, Hawaii; NSO/Sacramento Peak Observatory, in New Mexico; Observatorio Astronómico Nacional, in San Pedro Martir, Baja California, Mexico; and Panguitch Lake, in Utah. Each has a small coronagraph and monitors of dust, water vapor, and other atmospheric conditions. The quality of the seeing is also being studied.

The current schedule is to choose the site in 2004 and to construct the telescope between 2006 and 2009. The completion date may be much later, though, depending on how the funding develops.

The Least You Need to Know

- ◆ A Canary Islands telescope gives images of the highest resolution.

- ◆ Solar telescopes in India and elsewhere track and study the Sun.

- ◆ Coordinated sets of radio telescopes can image the Sun in radio waves.

- ◆ The site is being selected for a huge solar telescope that will use the latest technology.

20

Ringing Like a Bell

In This Chapter

- ◆ Waves galore
- ◆ Sunquakes
- ◆ Seeing around the back
- ◆ The world is round

Advances in astronomy often follow technology. Things that were discovered in small quantities and with great difficulty using older techniques can now be studied in large quantities and, though still with difficulty, straight-forwardly. The study of waves on the Sun follows this pattern. Discoveries made decades ago with film turn out to have widespread and major impor-tance. They are now followed minute to minute with electronic detectors.

Images, Plus or Minus

Images have been made on film for over a century, but those images are relatively difficult to manipulate. Nowadays, more people are using scanners and Adobe Photoshop or similar programs to manipulate the images they take. But in the 1960s, it took some ingenious work at Caltech to make the significant manipulations that led to important discoveries around the Sun.

Robert Leighton, Robert Noyes, and George Simon realized that parts of the solar surface went up and down with a period of five minutes. They had discovered a "five-minute oscillation."

For years, the five-minute oscillation seemed like a curiosity. Theoreticians worked on understanding the sizes and motions on the Sun that could lead to such a period. But at first there was no inkling that the oscillating motions on the Sun were a widespread phenomenon.

For one thing, it is hard to study the Sun for extended periods of time. The seeing varies, for example, and images taken under poor conditions of seeing are hard to compare with images taken under good conditions. Furthermore, the Sun sets at night, so it is difficult at a given site to take images for more than 12 hours or so (somewhat longer in the summer). And telescopes differ enough from each other that it is usually difficult to compare in great detail images taken with different telescopes.

 Solar Scribblings

An ingenious way of getting long periods of solar observations is to go near one of Earth's poles, where the Sun never sets for months on end. An early attempt was in Thule, Greenland, where a run of 60 consecutive hours of sunlight was obtained before clouds moved in. Later, long runs of solar observations were taken with a telescope at the South Pole.

Different techniques have been used to measure velocities on the Sun. The most accurate way is to use a spectrograph to measure a spectrum. The spectral lines are shifted to the blue when gas on the Sun is approaching, and shifted to the red when gas on the Sun is receding. But these shifts are relatively small for common motions on the Sun, so accurate techniques were needed to measure them. Use of an iodine cell, an iodine-filled container with transparent ends, enabled accurate measurements of motions.

Analysis of the long runs of data revealed that waves of many different periods were occurring on the Sun. Not only were there five-minute oscillations, but there also were oscillations many hours or days long. The longer the trains of uninterrupted data were, the more it was possible to detect the oscillations of longer periods. It became clear that major projects should be started to obtain solar velocity data all the time.

Fun Sun Facts

Although iodine is an excellent antiseptic, something to put on your finger when you cut it, it turns out to have been key in discovering more than 100 new worlds. The vapor of iodine, one of the chemical elements, has an absorption spectrum of very sharp spectral lines. When a tube containing iodine vapor is placed in the outgoing beam of starlight from a star under study, the vapor superimposes sharp absorptions at known wavelengths. These sharp iodine absorptions allow the wavelengths of absorption lines from the star itself to be measured to the fantastic accuracy of 3 m (10 feet) per second. Measurements to this accuracy have led to the discovery that the spectral lines from over 100 stars are moving in wavelength in a periodic fashion, revealing that the stars are being orbited by planets. A thorium comparison spectrum, as of 2003, is providing even better accuracy, to 1 m/s (3 feet/s).

Devastation on the Richter Scale

Waves that travel through Earth are known as *seismic waves*. Scientists measure the earthquake waves with devices called *seismometers*. Seismic waves spread through Earth and can be measured at various distances. Seismometers at different locations measure the waves they receive from an earthquake; by triangulating, scientists can figure out where the earthquake was and how strong it was.

Waves that spread through Earth can be useful, especially when they are not strong enough to cause damage. Changes in the speed at which the waves propagate reveal conditions under Earth's surface, where we can't see directly. So *seismology* on Earth, the study of seismic waves, is used to map out Earth's interior. By analogy to terrestrial seismology, the use of waves on the Sun to map out the Sun's interior is known as *helioseismology*. Waves on the Sun move at speeds of about half a kilometer per second. A full wave may let part of the Sun rise and fall over a distance of about 40 km (25 miles).

Sun Words _____

Seismic waves are vibrations that travel through Earth.

Seismometers are devices used to measure seismic waves on Earth.

The study of seismic waves is known as **seismology.**

The study of waves on the Sun and their use in mapping the solar interior is known as **helioseismology.**

Solar Scribblings _____

Chinese scientists thousands of years ago invented a simple seismometer in which balls rested insecurely on surfaces. When the ground shook, the balls fell off. Today's seismometers still use objects that are free to shake when the ground shakes.

In and Out

When I was growing up in New York, kids traditionally played with pink rubber balls called Spaldeens. (Their actual name was Spalding, but don't try to tell that to New York kids who grew up in a certain era.) When you were waiting to bat or throw, you could squeeze the ball in your hand. The difference between squeezing balls side to side and both side to side and top to bottom simultaneously illustrates differences between the ways that the Sun oscillates. Helioseismology studies the types of ups and downs of the solar surface and the large-scale patterns they have.

> ## Solar Scribblings
>
> Planets and their moons are squeezed in different directions. The most prominent example is Jupiter's moon Io, which is squeezed this way and that by gravity from Jupiter and the other major moons. The squeezing has heated up Io's interior, leading to extensive volcanism. But the Sun is heated so much by nuclear fusion that any effect of squeezing is negligible for heating, and the solar oscillations are generated by waves inside the Sun rather than by outside sources.

Solar seismologists categorize the waves by how many wavelengths it takes to fit once around the Sun. Since the Sun rotates, the direction around the equator is spinning and things relative to the equator can be different from measurements perpendicular to the equator. So you can speak of a type of wave that has a wavelength of, say, 50 times around the Sun's equator but only 10 times from pole to pole. Each type of wave can be assigned numbers for these two types of wavelengths. Note that we are talking about the Sun's whole surface moving in and out, with areas moving inward adjacent to areas that are simultaneously moving outward. A whole lot of different types of waves are all going on at the same time.

The surface oscillations are caused by sound waves generated by the convection that we see on the surface as photospheric granules. These waves are trapped inside the Sun, setting the solar surface vibrating. Since sound waves are variations in pressure, these waves are known as p-modes. As the waves in the solar interior move inward, they are bent by the changing speed of sound, which results from the rising temperature. As a result of the bending, the waves wind up hitting the surface. The rapidly decreasing temperature and density there cause the wave to be reflected back downward.

A set of images of part of the Sun made at Caltech in the 1960s as part of the work that led to the discovery of the five-minute oscillations. The images at the right show the velocities that result from subtractions of pairs of images on the left. Parts a, c, e, g, and i show redshifted gas and b, d, f, and h show non-moving gas. Subtracting pairs of these gives the images on the right, which show velocity. Parts a' and g', taken five minutes apart, look more alike than other parts, revealing a five-minute oscillation.

(A. Title, Lockheed Martin Advanced Technology Center)

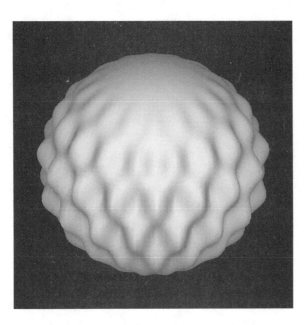

One of the many modes of oscillation, with the extent of the excursion of the solar surface magnified over 1,000 times.

(NASA's Marshall Space Flight Center)

Solar Scribblings

When you tune your radio, you are picking just one set of wavelengths from the air. All the different wavelengths are around us all the time, with the radio spectrum invisible to us. But with the right electronics, we can pick up any given station. Similarly, the Sun has thousands upon thousands of different wavelengths going on all the time, and the task of helioseismologists is to pick out individual ones for study from the cacophony of what is always going on.

Depending on the wavelength, waves penetrate into the solar interior for different distances. Short-period waves, like those of the five-minute oscillation, are limited to regions near the Sun's surface. Waves of periods of hours go much deeper. Therefore, it is desirable to study the longest-period waves in order to find out about the Sun's deep interior.

The Solar Scoop

The longest-period waves have taken us to the deep interior, close to the region where nuclear fusion is fueling the Sun. They reveal the temperature in this region. Knowing the temperature is important for predicting the rate at which the nuclear reactions go on. In turn, that rate tells us the number of neutrinos that should be emitted from the Sun each second. Since we know our distance from the Sun, we can easily transform that rate into the number of neutrinos that should pass each square meter of Earth each second. As we have seen, this predicted number differs by a factor of about 3 from the number observed. This solar neutrino problem has recently been solved by the realization and proof that two thirds of the neutrinos are transforming themselves into other, previously undetected types.

Results from helioseismology have made our knowledge of the inside of the Sun more accurate. For example, helioseismology has told us that the outer 30 percent of the Sun is convective, with hot elements of gas rising like the hot bubbles of boiling water. Helioseismology has revealed how fast the different levels and latitudes of the Sun's interior are rotating. Knowledge of the rotation is important for understanding the formation of sunspots and solar activity that impacts Earth.

What's Behind?

When we look at the Sun, we see only half of it at a time. The other half is around the back. Since the Sun rotates about once a month, a bit of the surface on one side has just disappeared, but the part of the surface that is about to rotate into view hasn't been seen for about two weeks. It could have changed drastically during that time. Maybe an active region is about to appear and shoot off lethal particles at us on Earth.

Sometimes loops of gas stick up off the solar surface far enough that we can see them a day or so in advance, even though their base is on the far side. But we cannot directly see active regions farther around on the back, even though particles that they eject into space can curve around to hit us.

Helioseismology provides a way to get at least a rough view of what might be on the Sun's far side. Waves from far-side active regions travel through the Sun and affect oscillations that we see on the side of the Sun facing us. In essence, these waves make the Sun transparent.

To make the necessary calculations to trace back oscillations of our side of the Sun, we need very detailed observations. These have been provided by the MDI instrument aboard NASA and the European Space Agency's SOHO spacecraft, which we discuss in more detail in the next chapter. MDI routinely measures a million points on the side of the Sun facing us. After the waves are released by a sunspot region on the far side, the strong magnetic field in that region speeds up the waves passing through it. These waves pass through the Sun, as a result, about 12 seconds faster than other waves, compared with the overall travel time of about 6 hours.

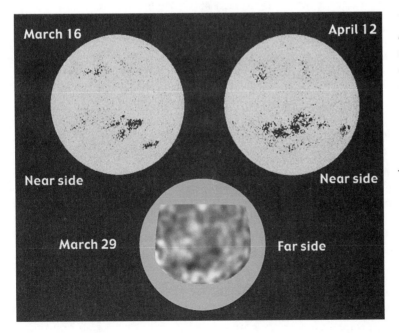

The blurry image at lower center is an active region on the Sun's far side and is otherwise invisible except for this helioseismology method. This region is at the center of the Sun's disk two weeks before (upper left) and two weeks after (upper right) its far-side imaging, and the images show how rapidly it changes.

(Charles Lindsey, Solar Physics Research Corp., and Douglas Braun, NorthWest Research Associates)

'Round the World in Six Stops

As you learned in Chapter 3, the U.S. National Solar Observatory has set up a Global Oscillation Network Group, GONG. It got its name because it is studying how the Sun is ringing like a bell or a gong. (Rather than ringing from a single note and from a single strike, though, the Sun's ringing is more like linked bells hit by a sandstorm.) A bell or gong gives off sound waves, and the Sun's surface vibrations show the presence of similar waves there. From helioseismology, some of the waves are more significant than others. Studying all the waves together tells us about the solar interior, but it is important to get the longest possible complete runs.

The U.S. National Solar Observatory heads the GONG project. Six telescopes spread around the world give coverage over 90 percent of the time. Each has an identical set of instruments and is fed by identical telescopes. For some of the wave studies, all the light from the Sun is considered together, without resolving individual surface features. The GONG telescope is only 8 cm (3 inches) across. The instruments record solar data every minute.

At each site, a pair of mirrors reflects the sunlight into a trailer, which houses all the instruments. Construction of identical trailers took place in the United States, after which the complete instrumental setups were shipped to the sites all over the world.

The six sites, all in operation for over half a dozen years, are listed here:

- Big Bear Solar Observatory, in California, USA
- High Altitude Observatory, at Mauna Loa in Hawaii, USA
- Learmonth Solar Observatory, in Western Australia
- Udaipur Solar Observatory, in India
- Observatorio del Teide on Tenerife, in the Canary Islands
- Cerro Tololo Interamerican Observatory, in Chile

Electronic capabilities have advanced in the half-dozen years since deployment of instruments at the GONG sites, and in 2001 upgrades transformed the setups into what is called GONG+. GONG+ uses electronic detectors (CCDs) that are 1,024 pixels square—a total of over a million pixels. These detectors give pixels that each correspond to 2.5 arc seconds on a side when viewing the Sun, allowing features about 5 arc seconds across to be imaged. Seeing (image quality) is not much better than that, typically, so it was not thought useful to provide a larger telescope or make other modifications.

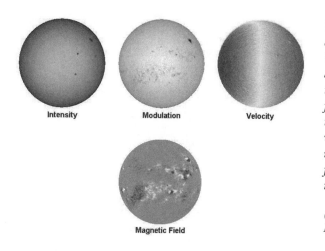

Intensity   Modulation   Velocity

Magnetic Field

The different kinds of solar data available from GONG++. We see the overall intensity that is observed; the variations of brightness from point to point (modulation) after various slowly varying backgrounds are removed; the overall velocity field, reflecting the sun's rotation; and the magnetic field.

(David Hathaway, NASA's Marshall Space Flight Center)

Helioseismology has provided us with an astonishing number of parameters about the solar interior. They include the ranges and position maps of temperature, density, the speed of sound, and the velocity of rotation. With GONG++, we will now have continuous measurements of the changing flows beneath the surface, that we call "solar subsurface weather." We have also learned the depth of the boundary of the convection zone, the total abundance of the elements heavier than hydrogen compared with that of hydrogen, and the abundance of helium, the second-heaviest element and a key to cosmology. Though it wasn't obvious until it was discovered, helioseismology is an extremely fruitful method of studying the Sun. It is now being applied as well to more distant stars, though to a lower level of accuracy because the stars are so much fainter.

The Least You Need to Know

◆ The Sun's surface is bouncing up and down, ringing like a bell.

◆ Analyzing the surface motions reveals the conditions of the Sun's interior.

◆ Waves on the near side reveal even sunspots on the Sun's far side.

◆ A network of solar telescopes spread around the world gives nearly continuous coverage.

Part 5

The Sun from Space

If something is in your way, get around it. To study the Sun, our Earth's atmosphere is the biggest barrier. With the space age, we can get above the atmosphere to watch the Sun closely and continuously. As we get more capable, we see increasingly finer detail, revealing gorgeous looping gas racing around just above the Sun's surface. And we even have unfinished plans to go up close to the Sun—though we are unable to do it at night in order to keep cool.

21

Above the Air Is Better

In This Chapter

- ◆ Rockets take us above the atmosphere to study the Sun
- ◆ NASA has launched a series of Orbiting Solar Observatories
- ◆ Crewed spacecraft have studied the Sun
- ◆ Views from over the solar poles are rare and reveal the solar wind

One of the popular exhibits at the Smithsonian's National Air and Space Museum in Washington, D.C., is the backup module to the one that astronauts used on NASA's Skylab Mission. This space station was an early attempt to make scientific observations from space, and the Sun was at the center of the astronauts' attention. The Sun has been the subject of study since the beginning of the space age.

The Air Is Blocking Your View

Several technological developments from World War II proved useful to astronomy after the war ended in 1945. The Sun, as the nearest and brightest celestial object, was the first focus of attention.

The Sun and Radar

Radar was developed especially by American and British scientists in the 1930s and 1940s, following some earlier roots. Radar, which stands for radio detection and ranging, involved the development of electronic devices to detect reflected radio signals sensitively. For a time, it was a military secret that the Sun was a strong emitter of radio waves, for it blocked radio observations when it was active. For example, all radio communications were blocked in England on a range of radio bands for two days in February 1942. Everyone's first thought was that the Germans had worked out a way of blocking radar, but then James Hey realized that the strongest static began when the Sun rose and ended when the Sun set. Though he connected this static with a large sunspot group that was on the Sun, his reports were kept secret until the war was over. An American radio scientist also figured out that the Sun was giving off radio waves, yet couldn't publish until the war was over.

During and after the war, almost nobody thought that the Sun would be significant for astronomy, and it took an amateur astronomer, Grote Reber, to use radio waves to map the sky. He found various radio sources other than the Sun and wrote a scientific article announcing solar radio radiation in 1944. His article was the first to be published that announced the Sun as a radio source.

Other scientists later tracked several kinds of radio bursts. Understanding and predicting them had obvious applications to the security and availability of communications. But radio waves come through a window of transparency down to Earth's surface every day, so they are not the subject of this chapter.

High Above the Atmosphere

The Earth's atmosphere blocks out all the radiation shorter than visible light. In particular, ultraviolet, x-rays, and gamma rays do not penetrate through the atmosphere to the Earth's surface. Thus we must go up into space to study them. Observing the sky from space started with the V-2 rockets captured from Germany; perhaps a hundred of these rockets were brought back to the United States. Rocket scientist Wernher von Braun and most of his team came with them. In the late 1940s, American scientists began using the rockets. The military background of the rockets explains in part why scientists at the U.S. Naval Research Laboratory were pioneers in using them.

The rockets stay aloft for about a half hour, of which about five minutes is prime viewing time. A door on the rocket opens, and a telescope or camera points at the Sun, so many things must happen quickly and many things can go wrong. But even though fewer than half the rockets succeeded, the scientists obtained ultraviolet spectra of the

Sun. These spectral lines revealed conditions in the upper levels of the Sun's atmosphere. At first, the resolution of the spectral lines was poor and the spectra did not go very far into the ultraviolet. But gradually, the quality of the observations improved. Through the late 1950s, many images and spectra were obtained of the solar x-rays and ultraviolet. The data backed the idea that the solar corona is very hot and gave a variety of details. Furthermore, the capabilities left American astronomers ready to plunge ahead when satellites were first launched.

Science was transformed and the space age began on October 4, 1957, when the Soviet Union launched Sputnik. Though Sputnik itself was not a research satellite, its existence galvanized the scientific establishment in the United States and elsewhere. Not coincidentally, 1957–1958 was the 18-month International Geophysical Year (IGY), when scientists all around the world teamed up to concentrate on getting data about the Sun and its relation to the Earth. Preparing for the IGY led to the growth of solar astronomy and solar observatories.

Soon thereafter, NASA began a series of solar observatories in space, beginning with Orbiting Solar Observatory 1 (OSO-1) in 1959. Gradually, the reliability of the spacecraft and the quality of the observations increased. For example, by OSO-6, the resolution obtained was much improved; OSO-8 carried cameras capable of high-speed observations of spectral details. The Soviet Union also sent some spacecraft aloft to make solar observations, though not with such a series of vehicles.

Solar Scribblings

Computers were much more rudimentary when the OSO projects were the latest thing, and there was little imaging that computers were capable of doing. When OSO-4 data came down to Earth, we at the Harvard College Observatory had printouts of numbers showing the strength of intensity from various regions on the Sun. We sat down with markers to color in the regions of similar intensities. I remarked then that it was good to be using the skills I had mastered in kindergarten. The senior professors joined us students in the coloring.

People Aloft

The current methods of lofting rockets are very wasteful. A huge tank of rocket fuel goes aloft with a smaller rocket carrying the payload attached. The Apollo program was devoted to bringing people to the Moon safely and returning them to Earth. It brought 12 astronauts on 6 missions to land on the Moon during 1969–1972. But then, though three additional Apollo missions to the Moon had been planned, they were cut for lack of funds. The program was perhaps a victim of its own success, since the last several Apollo missions had brought back so many moon rocks and sent back so much information that funders (not scientists!) deemed future missions unnecessary. Apollo 17 from 1972 remains the last time that people went to the Moon.

But NASA had some additional giant Saturn V rockets. They devised a space laboratory to go into Earth orbit, using the empty fuel tank of one of the Saturn V rockets as a laboratory. Attached to that Skylab was the Apollo Telescope Mount. This set of telescopes and instruments was largely devoted to solar observations. Skylab was launched in 1973.

Solar Scribblings

Skylab was not launched very successfully. A shield meant to protect the mission from micrometeoroids broke lose and tore off one of the giant solar panels meant to provide power. The first crew of astronauts to visit Skylab had its repair as their first task. They devised an umbrella to replace the shield, which also kept Skylab usably cool by blocking sunlight from hitting it directly. Over the next eight months, three crews of astronauts successfully used the Apollo Telescope Mount.

Skylab, with its remaining solar panel and jury-rigged sunshade. It carried several solar telescopes on its Apollo Telescope Mount.

(NASA's JSC)

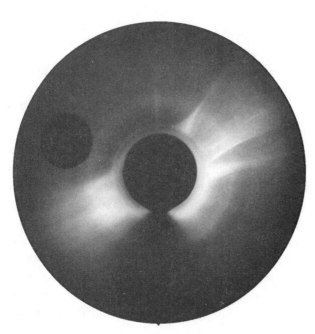

An image with Skylab's coronagraph. Note the size of the Moon (left), which was near to but not in the field of view; a total solar eclipse was visible from Earth that day.

(NASA)

The Apollo Telescope Mount's instruments included two devices to study the solar ultraviolet. It also included a coronagraph to observe the outer corona in visible light. And it had x-ray telescopes that gave the finest x-ray images yet available. The resolution of the x-ray images was so fine that it became clear how dependent the heating of the solar corona was on the Sun's magnetic field; the corona was hottest over the high-magnetic-field sunspot regions. These observations led to the quick demise of the old theory that sound waves from below the photosphere heated the corona.

Solar Scribblings

One of the Apollo Telescope Mount's x-ray telescopes was built by the American Science and Engineering Company in Cambridge, Massachusetts. Soon thereafter, the whole scientific group moved over to the Harvard-Smithsonian Center for Astrophysics. Its head, Riccardo Giacconi, went on to make many discoveries in x-ray astronomy, for which he received the Nobel Prize in Physics in 2002. Giacconi had earlier become head of the Space Telescope Science Institute and then of the European Southern Observatory. American Science and Engineering retained its expertise in x-ray work and is a major provider of x-ray machines used at security checkpoints in airports.

Solar to the Max

The next major solar spacecraft was timed to coincide with a maximum of the solar-activity cycle. It was therefore especially well situated to study flares and other activity. This Solar Maximum Mission (SMM) was launched in 1984. It carried instruments from both American and European institutions.

Among the devices that SMM carried aloft was a coronagraph that was capable of more detailed images than earlier space coronagraphs. Another interesting instrument measured with great precision how much energy was coming from the Sun day after day and how it changed over time. We discuss this solar constant and its measurement in Chapter 26. Suffice it to say here that it was proved that the solar constant wasn't really constant. Furthermore, though the solar constant declined for a while, it was clear that the rate was too high to continue, since the Sun would then waste away quickly. Still, there was some relief when the solar constant began going back up, showing that it kept in phase with the solar-activity cycle. The instrument also showed that the presence of a large sunspot on the Sun reduced the total solar radiation measurably.

> ### Solar Scribblings
>
> I had a personal interest in the measurements of the solar constant by Solar Maximum Mission. On the oral exam for my Ph.D. some years earlier, I had been asked what effect the presence of a large sunspot had on the Sun's total emission. I hemmed and hawed, since I didn't know the answer and I could argue it theoretically either way. It was possible that the sunspot just blocked enough radiation to be measurable. But it was also possible that the radiation that didn't come through the sunspot heated adjacent areas and that the total amount of emission was the same. Nobody told me that they had asked a question that had no known answer! Anyway, I passed.

For its work at wavelengths shorter than the visible, Solar Maximum Mission carried a gamma-ray spectrometer, three x-ray telescopes that were able to measure spectra, and a telescope to measure polarization and spectra in the ultraviolet. Its coronagraph, for use in a wide range of visible light, concentrated on one quadrant of the Sun at a time. It also carried a device to measure the solar constant, the total amount of energy emitted by the Sun each second as received by a square centimeter at the top of the Earth's atmosphere. We discuss results from this Active Cavity Radiometer Irradiance Monitor (ACRIM) in Chapter 26.

Although SMM gave excellent results for many months, nine months after launch it started spinning uncontrollably. The spacecraft seemed entirely out of touch. NASA wanted to show that its astronauts could fix things in space, perhaps as a precursor to its efforts in building crewed space stations and in part to show the worth of the space-shuttle program. A crew of astronauts went aloft in 1984 and succeeded in bringing SMM into the space shuttle's cargo bay, where they fixed it. When they dropped it off, it lasted another five years.

> ### Fun Sun Facts
>
> Solar Maximum Mission was the last major mission that NASA put into space using single-use rockets. The subsequent larger missions have been launched from space shuttles, though several smaller missions were launched from rockets.

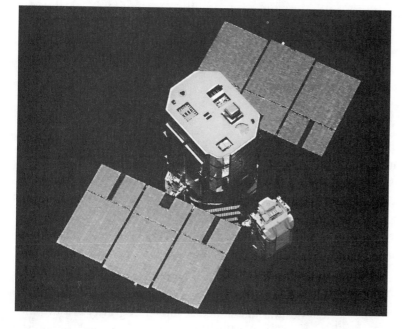

An astronaut attached himself to SolarMax to repair it.

(NASA's JSC)

Spacelab

At its inception, the space-shuttle program was promoted as a cheap way to get things into space. Unfortunately, it hasn't worked out that way. And when the *Challenger* space shuttle exploded in 1986, killing the astronauts aboard, our view of the program as an inexpensive Space Transportation System, trucking objects safely into space, was transformed. Was it really worth using humans to launch systems that could have been carried aloft by robotic spacecraft? The question was compounded by the February 1, 2003, break-up of the space shuttle *Columbia* as it re-entered Earth's atmosphere, killing the seven astronauts. As of this writing, with investigations of the *Columbia* disaster underway, it is impossible to foresee the future of the crewed space program or its shuttle component.

The space shuttles were at first held out as homes for space laboratories. A Spacelab module was prepared, and it was hoped that it would be launched many times. As it turned out, it was used only a couple of times. Solar instruments were aboard Spacelab 1 when it spent 11 days in orbit in 1983. Spacelab 2 was aloft for a week in 1985. Very high-resolution images were obtained of fine structure on the solar surface with the Solar Optical Universal Polarimeter (SOUP). Because these observations were steady in quality, free of variations caused by seeing effects in Earth's atmosphere, the evolution of such small regions of the photosphere could be followed. Both images and magnetic-field measurements were obtained so that the two could be correlated. Other instruments were devoted to solar ultraviolet studies. Spacelab 3 was also aloft in 1985—actually three months before Spacelab 2—and took solar spectra. In all these cases, many other instruments were devoted to other fields, such as life sciences and materials science.

The Voyage of Ulysses

In the *Odyssey* by Homer, the Greek hero Ulysses journeys for many years on his way home from the Trojan wars. The spacecraft Ulysses is undergoing an even longer journey through the solar system. It was launched in 1990 from a space shuttle, but it headed outward in the solar system even though it eventually would go toward the Sun. All previous spacecraft had been in the plane of the Earth's orbit, which corresponds roughly to the Sun's equator. Thus, we never get a good look at the Sun's poles, since the Sun is tipped only slightly.

Fun Sun Facts

Ulysses was originally supposed to be a pair of spacecraft, known as the Solar Polar Mission. But a funding crisis led the United States to back out, angering its European partners. Eventually, one European spacecraft was launched by NASA, making it a joint NASA/ESA mission.

But getting high over the Sun's poles would be very expensive in terms of fuel if we went there directly. So Ulysses was first aimed at Jupiter. A gravity assist from Jupiter—almost like bouncing off it, though the only impact was through gravity—then sent Ulysses high over the Sun's south pole. After passing over the south pole in 1994, Ulysses curved back and flew over the Sun's north pole in 1995. It next went back to the outer solar system, only to return to the Sun again. It then flew over the Sun's south pole in 2000 and the north pole in 2001. Ulysses's main task was to study the solar wind, the outflow of the solar corona.

For reasons of funding, Ulysses didn't carry a camera, so we don't have images looking down at the Sun's poles. But Ulysses is carrying instruments to measure particles that hit it and to measure the magnetic fields it encounters.

Ulysses's unique perspective led to many discoveries. For example, it sampled a huge bubble in the solar wind.

Ulysses is funded until late 2004, allowing it to complete a whole sunspot cycle, though it is at a different location at each time, making it difficult to disentangle what variations come from changes in the Sun and what variations come from the spacecraft's varying location.

The Sun's north and south magnetic poles interchanged at the peak of the solar-activity cycle, which occurred in 2001–2002. So the remaining time for Ulysses will allow it to see how the heliosphere settles down following the major change in magnetism. One way the heliosphere will be studied is by observing the Sun's magnetic field. When the field is strong, it bends cosmic rays from beyond the Sun, deflecting them away from the Earth. So cosmic-ray sampling will be a key to understanding the changes in the heliosphere.

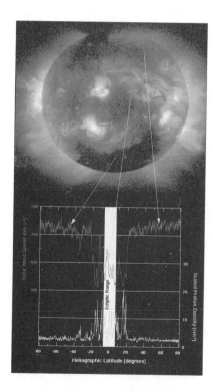

The graph shows the speed of the solar wind (top curve) and the density of gas (bottom curve) as Ulysses went from over the Sun's south pole to over its north pole. Before Ulysses, only the region marked with the vertical bar in the middle had been sampled.

(LANL, NASA/JPL, and ISAS)

The Least You Need to Know

◆ Early solar astronomy used rockets and satellites to get above Earth's atmosphere.

◆ Rockets and spacecraft study especially the ultraviolet and x-rays, with some visible coronagraphic work.

◆ Astronauts have handled and repaired solar spacecraft.

◆ With an assist from Jupiter's gravity, Ulysses went over the Sun's poles.

Sunbeam

In This Chapter

- ◆ X-rays penetrate
- ◆ Rockets go high enough to record x-rays
- ◆ Yohkoh mapped the x-ray Sun for 10 years
- ◆ Other spacecraft also map high-energy solar radiation
- ◆ X-ray telescopes can piggyback on Earth-viewing spacecraft

As the space age advanced, glimpses of the Sun from rockets gave way to spacecraft that provided longer views. During the 1990s, a Japanese spacecraft named Yohkoh provided continuous, high-quality x-ray images that allowed scientists to study violent processes on the Sun.

Imaging X-Rays

When Wilhelm Roentgen discovered x-rays in the 1890s, he soon produced a photograph of his wife's hand, with her wedding ring showing prominently around the finger bone. Images like that dazzled the world. Such x-ray images show both the desirability of using x-rays and some of the problems with them.

Roentgen's wife's hand, including her ring, taken in 1895. This image is often called the first x-ray photo.

(Radiology Centennial, Inc.)

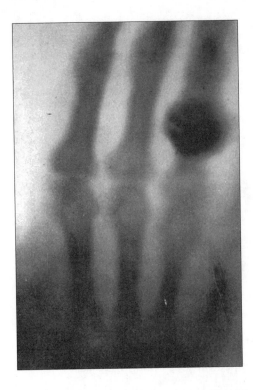

The x-radiation is produced by high-energy processes. Your doctor or dentist's office has such devices in order to produce the x-rays used for diagnosis. But we are not x-raying the Sun by putting radiation through it. Rather, we are examining directly the x-rays that are generated by high-energy and high-temperature processes on the Sun.

Solar Scribblings

Why are they called x-rays? Because they were mysterious. Not long after Roentgen found them, many other types of rays were reported. One, in particular, was called N-rays. Dozens of other scientists confirmed the detection. But the existence of N-rays seemed implausible on other grounds. In 1904, a scientist investigating the N-rays secretly removed the prism that was reported to be necessary to produce them; the scientist who initially reported the N-rays claimed to see their effects anyway. His self-deception was thus exposed. The episode drove the original scientist mad.

The hotter a gas is, the farther to shorter wavelengths its radiation peaks. That is, there is some wavelength at which a gas gives off more energy than at either longer or shorter wavelengths. The position of this peak wavelength shifts to shorter wavelengths as the gas gets hotter. For the solar photosphere, with a temperature of about 6,000 kelvins, the peak is in the yellow-green part of the visible spectrum. For the solar corona, with a temperature of perhaps 3,000,000 kelvins, the temperature is 500 times higher, so the peak of the wavelength is 500 times shorter. That puts the peak in the x-ray region of the spectrum.

You can't merely put an ordinary lens in a telescope to focus x-rays, since the lens would just get x-rayed. Scientists use several techniques to make images of celestial objects with x-rays. One way is to use a set of holes or channels, much like the honeycomb of metal that you often see on commercial fluorescent light fixtures. Only the x-rays going straight ahead through the small channels are recorded, and their position on film or electronic sensors can be noted. Especially for the longer-wavelength part of the x-ray spectrum, thin coats of metal can be deposited on glass lenses. The x-rays bounce back and forth between the coats of metal, and the property of light known as interference allows images to be made.

The property of grazing incidence can be used to focus all x-rays, including the shortest. If you throw a rock straight down on a body of water, it may just make a plop and not bounce off. But if you throw a rock across the same body of water at a low angle, it may skip across at a low, glancing angle. The same property is used to make high-resolution x-ray telescopes. NASA's giant Chandra X-ray Observatory, the x-ray equivalent to the Hubble Space Telescope, was launched in 1999 with a nested set of cylindrical mirrors, each of which focuses x-rays at glancing angles. Because the mirrors are intercepting the incoming x-rays almost edge on, they present a very small surface for collecting the radiation, which is why the mirrors are bent around in cylinders with several of them nested together—just to provide more area.

The Solar Scoop

The Chandra X-ray Observatory uses grazing incidence to make very high-resolution images of celestial objects. Its resolution reaches half an arc second. But the Sun is too bright a source to point directly at; the intense sunlight would burn out the detectors.

Solar Rockets

Though spacecraft are best for long-term monitoring of the Sun, there is still a substantial advantage to using rockets. For one thing, once the spacecraft are aloft, we usually can't modify them. Their sensitivities can change, for example, fooling scientists on the ground. Furthermore, sounding rockets are cheaper than orbiting spacecraft.

It is thus useful to launch, from time to time, a rocket that carries well-calibrated instruments. In the five-minute period or so that it is above the atmosphere, it can make observations that overlap the satellite observations. When the rocket is recovered back on Earth, its instruments can be checked to verify that their sensitivities haven't changed. Then the rocket results can be compared with the spacecraft results to provide accurate calibration of the latter.

Furthermore, the rockets can carry instruments or filters that aren't on the existing spacecraft. For example, a rocket timed to coincide with the 2001 eclipse made observations in 16-times-ionized nickel, a type of radiation released at higher temperatures than could be measured from the spacecraft then aloft. A rocket launched in 2002, whose results were reported at a 2003 meeting of the American Astronomical Society, also provided unique data.

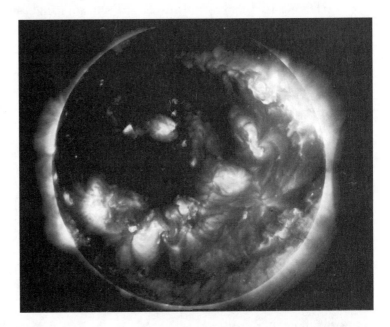

The x-ray corona in as much detail as has ever been seen, imaged with a rocket in soft x-rays using face-on rather than grazing incidence techniques. The flight took place on the same day that a total solar eclipse was visible from Earth. See Chapter 9 for another rocket x-ray solar image.

(Leon Golub, Smithsonian Astrophysical Observatory)

Fun Sun Facts

An eclipse shadow moves across the surface of the Earth at speeds that get as low as 1,000 or 2,000 km/hr. But orbiting spacecraft usually move at 18,000 miles per hour. So the mismatch between a spacecraft and an eclipse is worse than the mismatch between the ground and an eclipse. Even if the spacecraft passed through the umbra, it would do so in seconds. So the purpose of eclipse rockets is to view the Sun within hours of the eclipse so that the measurements can be compared with those made during totality. Indeed, x-ray observations from the rockets show the corona across the face of the Sun, which would be hidden if the rockets were themselves inside the zone of totality.

Yohkoh's Decade

After decades of progress in solar x-ray astronomy, it became possible to have an x-ray telescope in orbit in space for an extended period. The Japanese space agency, the Institute of Space and Astronautical Science (ISAS), stepped up to the plate to launch a satellite to study high-energy phenomena on the Sun, largely through x-ray images.

Launching Yohkoh

Earlier, in 1981, ISAS had launched a smaller mission to study solar flares. It was called Hinotori, which means "fire-bird." After its success, Japanese scientists and engineers designed a larger mission, called then Solar-A, to observe flares and other activity on the Sun with higher resolution. It would resolve better not only spatial details of flares, but also their energy ranges. Furthermore, it would be sensitive enough to record the quiescent regions between the active regions.

Solar-A was successfully launched on August 30, 1991, from the Japanese space facility. It was then named Yohkoh, which means "sunbeam." It is in a circular orbit, circling Earth every 96 minutes.

Using Yohkoh

Many of the images displayed with discussions of Yohkoh were taken with the Soft X-ray Telescope, the United States' contribution to the mission. The telescope was built at the Lockheed Palo Alto Lab in collaboration with the National Astronomical Observatory of Japan, and was later controlled by scientists at Montana State University in Bozeman. It used a grazing-incidence telescope and a series of thin metallic filters to observe different wavelength bands between 3 and 12 angstroms. (Those wavelengths correspond to photons of energies between 4 and 1 kilo-electron volts [keV], the unit often used to measure energies.) The Soft X-ray Telescope had a resolution of 4 arc seconds, inferior to ground-based resolution in the visible part of the spectrum, but still very useful. It could take images every two seconds. Also important, it could record brightness over a very wide range, greater than a factor of 100,000. Thus, it could sense both faint objects and bright ones on the face of the Sun.

Yohkoh also carried a Hard X-ray Telescope. By "hard," scientists mean that the x-rays are more energetic, which means that they have shorter wavelengths. The Hard X-ray Telescope imaged flares with wavelengths from 5 to 20 times shorter than the Soft X-ray Telescope—that is, from 0.6 to 0.2 angstroms. The images had resolutions of 7 arc seconds. The telescope could take images as often as every half second, to follow the eruption and growth of a solar flare.

Yohkoh also carried devices to measure x-ray and gamma-ray spectra of flares.

The Death of Yohkoh

Yohkoh worked very well for about a decade, and a tenth-anniversary celebration was planned. Shortly before the date, Yohkoh passed into the eclipse path of what on the ground was the annular eclipse of December 10, 2002, observed in Costa Rica. For some reason, the sensors on Yohkoh went wild as Yokhoh passed into the eclipse path, and the spacecraft started spinning. In spite of hopes that it might be tamed, as Solar Maximum Mission had been 18 years earlier by astronauts and as the Solar and Heliospheric Observatory had been more recently using ground controls, Yohkoh never came back.

Yohkoh Science

The continuous stream of Yohkoh images allowed the overall structure of the Sun to be observed as it changed over the 11-year solar-activity cycle. The individual images show that the hottest regions of the corona, as revealed by their brighter x-ray emission, correspond to the regions of the strongest magnetic field in the solar photosphere. The magnetic field was mapped by ground-based observatories.

At the maximum of the sunspot cycle, in 1991 and in 2001–2002, not only are the magnetic field areas and sunspot areas larger than at all other times, but also the bright regions of the corona seen in x-rays are the most extended. The magnetic field lines that come out of the Sun broaden a bit by the time they reach coronal altitudes, so the x-ray bright regions are slightly larger in extent than the sunspots visible in the photosphere. At the minimum of the sunspot cycle, the overall x-ray corona is perhaps 100 times fainter and has many fewer magnetic regions showing. The bright regions that fade are replaced by only smaller, fainter regions.

After sunspot minimum, the orientation of the magnetic field changes. In particular, the positive polarity of the magnetic field leads the negative polarity in the northern hemisphere during one cycle, as the figure shows. Leading means that it goes first in the direction that the Sun is rotating. At the same time, the reverse is true in the southern hemisphere, with the positive polarity trailing the negative polarity. After sunspot minimum, these pairings are flipped. The positive polarity then trails in the northern hemisphere and leads in the southern hemisphere. Only after another full 11-year cycle do the polarities flip again and the Sun returns to its earlier magnetic configuration. For this reason, the true sunspot cycle is 22 years long, not merely the 11 years that you would superficially think, because you were looking only at the sunspot numbers rather than at the underlying magnetic field.

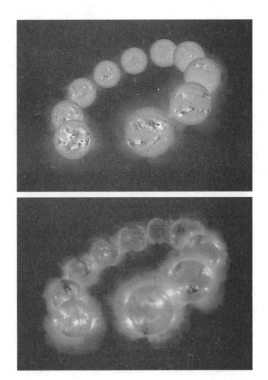

The variation of the solar magnetic field (top) and x-ray images from Yohkoh (bottom) over a whole solar activity cycle, in series of images taken approximately a year apart.

(X-ray images: LMSAL, Montana State U., NASA, NAOJ, U. Tokyo, and ISAS; magnetic images: NSO/AURA/NSF)

Exploring Small

NASA has several series of spacecraft in different price ranges. One of its Small Explorers is devoted to studying solar flares. Originally named the High Energy Solar Spectroscopic Imager, the spacecraft was renamed about a month after its launch in 2002 to honor the memory of the late gamma-ray NASA scientist Reuven Ramaty. It is in a low Earth orbit inclined 38°.

The Ramaty High Energy Solar Spectroscopic Imager (RHESSI) makes images with unprecedentedly high spatial and high time resolution, in order to study the explosive phases of flares and other solar phenomena. To make those images, it uses a metal grid; computers on the ground calculate the image that is built up as the spacecraft rotates. In hard x-rays, those at wavelengths less than 0.3 (20 keV in energy), its images have a resolution better than 4 arc seconds. Those images should improve to about 2 arc seconds as more information about the details of the spacecraft's operation are assimilated. RHESSI also measures the spectrum of those individual elements by recording the brightness in discrete energy bands. Furthermore, it takes spectra of gamma-ray spectral-line emission in flares.

RHESSI has observed thousands of flares in hard x-rays and dozens in gamma rays. The largest of these flares gives off more energy than a million megatons of TNT. The researchers are looking, in particular, for the locations of the emission of the most energetic x-rays and gamma rays, in order to find the sources of the huge amounts of energy. Scientists are sure that the energy is locked up in the magnetic field, but there is no agreement on the details of the mechanism for storing the energy or for triggering its explosive release. Spectra show that x-rays are generated from electrons given so much energy that they move at speeds greater than half that of light. These electrons stream downward along the magnetic field, starting at the peak of the coronal loop. The spectra also show that the gas is heated to tens of millions of degrees during the flare's peak.

Project scientists have localized powerful flares with RHESSI. By aligning these observations with those from other spacecraft, they have zeroed in on the flare mechanisms. The images have shown that the x-rays given off from the footpoints—the places where the flare loops descend that we see at higher levels are anchored in the photosphere—are caused by electrons slamming into the dense gas lower in the corona and in the chromosphere. It was a surprise that the x-rays were seen coming from the footpoints before other spacecraft showed brightening in their ultraviolet observations.

RHESSI's sensitivity is enough to find that tiny flares, lasting only a few minutes each, happen all the time. So x-rays are being generated at a low level fairly continuously—and not only in the few spectacular flares that previously were all that could be studied.

The solar flare of February 20, 2002, observed in different energy bands from the Reuven Ramaty High Energy Solar Spectroscopic Imager.

(NASA's Goddard Space Flight Center and UC Berkeley)

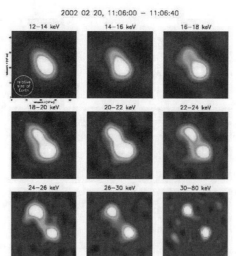

Go GOES

X-ray telescopes provide sensitive monitoring of solar activity. So it is useful to observe the Sun in x-rays constantly and to keep track of what is going on. But we don't always have to launch special spacecraft to do so. After all, we have many spacecraft aloft for other reasons. In particular, satellites in Earth orbit get their energy from solar panels. These panels always face the Sun, so why not just mount telescopes on them to study the Sun? This is now being done.

The Geostationary Operational Environmental Satellite (GOES) is a series of Earth-viewing satellites sponsored by the U.S. National Oceanic and Atmospheric Administration (NOAA). Satellites in this series send back pictures of the Earth's weather, among other things. As of the GOES-M, which was launched in 2001, an x-ray telescope was mounted on the solar panels.

GOES-M's x-ray telescope, which went into routine use in 2003, is a grazing-incidence model, in order to give high resolution that isn't available with other ways of focusing x-rays. It can send down x-ray images every minute, using several bands of x-ray wavelengths between 6 Å and 60 Å. NOAA scientists intend to use the image to study the relation of the Sun and Earth, and the images are immediately available on the web.

The Least You Need to Know

- X-ray telescopes take special forms to focus at such high energies.

- X-ray observations can be made from rockets aloft.

- The Yohkoh satellite made 10 years of x-ray solar images.

- A current spacecraft is observing x-rays and gamma rays from flares.

- Weather satellites can piggy-back x-ray solar imagers.

Chapter 23

Yo Ho, SOHO

In This Chapter

- ◆ SOHO carries a dozen instruments in orbit
- ◆ Several SOHO instruments study the corona
- ◆ SOHO provided a breakthrough in studying the Sun's insides
- ◆ Several SOHO instruments study the solar wind

Since Sputnik, many scientific instruments have been on satellites that orbit the Earth every 90 minutes or so. Furthermore, many weather satellites and other spacecraft are tens of thousands of miles high, appearing to hover over a given place on Earth. But a major spacecraft to study the Sun is in yet a different location: 1.5 million km (a million miles) away in the direction of the Sun. At that location, it remains fixed between the Earth and the Sun, retaining its excellent position for viewing. It also measures the influence of the Sun as it extends toward the Earth. For a decade, this spacecraft has been the best general observatory studying our nearest star.

Europe and America Conjoined

NASA and the European Space Agency don't always see eye to eye. They had a joint plan to launch a pair of spacecraft called Solar Polar Mission to fly over the Sun's poles, but the United States dropped out. Only the mission called Ulysses survived. On the Hubble Space Telescope, Europe is a junior partner, with 15 percent participation. But in the mission called the Solar and Heliospheric Observatory, the United States and Europe are really partners. If anything, as with Yohkoh, the United States is the junior partner.

> **Fun Sun Facts**
>
> The mass of SOHO's instruments is about 600 kg (1,300 pounds), less than that of a small car. Altogether, SOHO weighs a couple tons—or, at least, it did while on Earth before it was blasted into the weightlessness of space.

The Solar and Heliospheric Observatory, known as SOHO, carries a dozen instruments aloft. Of the individuals who head the different instrument teams, known as principal investigators, nine are European and three are American. Hundreds of scientists collaborate. The spacecraft was built in Europe and launched by NASA. NASA now receives the spacecraft data through its Deep Space Network of radio telescopes.

Previous solar observatories were in low earth orbit, which meant that the Earth blocked their view of the Sun for part of every 90-or-so–minute orbit. But full-time viewing is an advantage for many reasons, including the ability to obtain the long, continuous series of data that is useful for helioseismology. So SOHO was sent 1.5 million miles upward to a place known as the L1 Lagrangian point. There the net of the Earth's gravity and the Sun's gravity balance the centripetal force related to the speed and size of the orbit. As a result, SOHO can remain in place, orbiting the Sun with the same period as the Earth, while expending relatively little fuel.

> **Fun Sun Facts**
>
> SOHO isn't directly in line between the Earth and the Sun because, if it were, the Sun's radio emission would provide so much static that we couldn't get data from the spacecraft. Instead, it uses a bit of fuel to fly in a small circle around the point that is actually in line. The appearance of the small circle leads astronomers to say that SOHO is in a halo orbit.

SOHO was launched on December 2, 1995, from NASA's launch pad at Cape Canaveral, Florida. It was launched on an unmanned rocket. An attached Centaur rocket placed it into a circular orbit around the Earth. A second burn of the Centaur rocket then sent SOHO farther out into space, for the four-month trip toward its goal.

Dramatically, on June 25, 1998, a set of routine tests went wrong. Because of a human programming error, a computer program kicked in at the wrong time, and the spacecraft went spinning out of control. As SOHO lost power, the instruments and even the fuel chilled and everything stopped working. It took months before engineers on the ground could regain control, as the spacecraft's orbit changed enough for the solar panels to generate more electricity. Finally, and almost miraculously, all the instruments but one started up again. SOHO resumed its scientific mission and has worked flawlessly since then.

SOHO's Coronal Science

SOHO's instruments fall into separate categories. Some study the Sun's outer atmosphere. Some study the Sun's interior by observing the photosphere and using helioseismology. Others study the outflow of particles from the Sun known as the solar wind.

Since SOHO was launched at solar minimum and now has lasted through solar maximum, its instruments have given us a good view of how the Sun changes through the solar-activity cycle.

LASCO Blocks the Center

One of the simplest experiments on SOHO to follow involves a set of three telescopes. Each is a coronagraph. The inner one, now defunct, was an all-mirror coronagraph. Like traditional coronagraphs, it blocked the solar photosphere inside the telescope. However, it differed from the traditional design by not using lenses. The outer two LASCO telescopes, still operating well, are externally occulting coronagraphs. That is, a disk suspended in front of each telescope blocks the incoming light from the photosphere.

Each type of coronagraph—internally occulting and externally occulting—has its own set of problems. The innermost coronagraph, known as C1, had mirrors that were polished so carefully and so smoothly that they were known as superpolished. Nonetheless, the amount of light scattered in the coronagraph was enough to be troublesome. In any case, this was the only instrument that did not start working again after SOHO was saved from its spin-down.

Solar Scribblings

Astronomers have invented a robot to superpolish telescope mirrors. It monitors a mirror's surface to search for imperfections and programs the way in which the surface could be better smoothed. This technology may be applied to artificial knee joints: The lifetime of current artificial knees is limited by grinding in the joint, and superpolishing may help.

The other two coronagraphs show part of the corona very well, but they can't show the innermost part. The occulting disks in front of the telescopes are not in focus, since the telescopes must be focused at the Sun, which is millions of miles away. To properly hide the photosphere, the externally occulting coronagraphs must also hide the inner corona.

The C1 coronagraph showed the corona from 1.1 solar radii out to 1.5 solar radii. That is 1.5 times the radius of the Sun, which is almost 700,000 km (about half a million miles). The C2 coronagraph shows from about 1.5 solar radii out to 6.9 solar radii, while the C3 coronagraph shows from 3.5 solar radii out to 33 solar radii.

The Solar Scoop

Solar astronomers often measure distance into the corona in terms of the Sun's radius. If something is at 1.5 solar radii, it is half again the distance from the center of the Sun to its edge. This distance, the solar radius, is 700,000 km (420,000 miles). When the C3 coronagraph observes to 33 solar radii, it is seeing out around 22 million km (14 million miles) into space. This distance is a fair fraction of the 150 million km (93 million miles) of separation of the Sun from Earth.

The Sun seen with LASCO's second coronagraph, known as C2, which covers from about 1.5 to about 6.9 solar radii of corona.

(Naval Research Laboratory)

These coronagraphs have shown an astonishing phenomenon: Huge eruptions of mass occur about every day. As a result of this mission, these *coronal mass ejections* have been identified as a major source of the Sun's influence on the Earth. Watching the coronal mass ejections can even give warning hours or days before an eruption affects the Earth's magnetic field as well as satellites in orbit around the Earth.

Surprisingly, SOHO's coronagraphs have also led to the discovery of hundreds of comets. Many of these comets belong to a specific group of comets with very elongated orbits whose close point to the Sun is very close indeed. Sometimes these comets hit the Sun and disappear.

Sun Words

A **coronal mass ejection** is the eruption and departure from the Sun of a piece of the corona. These ejections, often called CMEs by professionals, have been seen near solar maximum to occur daily. They are a major link between the Sun and the Earth.

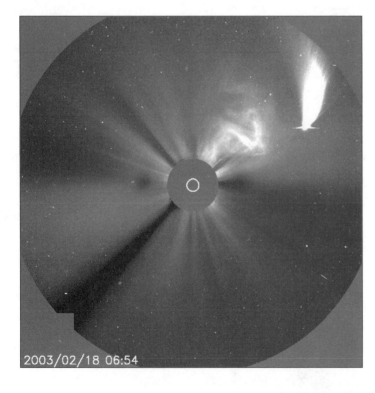

2003/02/18 06:54

A comet passing near the Sun, seen in LASCO's C3 coronagraph. The comet is C/2002 V1 (NEAT), the brightest comet seen by SOHO. The sun coincidentally gave off a coronal mass ejection during this part of the comet's closest approach. The overexposed comet nucleus caused the horizontal line of pixels that extend from it to appear bright. The size of the Sun is drawn onto the dark occulting disk for scale.

(Naval Research Laboratory and NASA's Goddard Space Flight Center)

Solar Scribblings

One of the first comets seen in a space coronagraph to disappear behind the occulting disk apparently hit the Sun. Soon thereafter, an apparent spray of material appeared from the other side of the occulting disk. People assumed that it was spray from the comet material. But now that we realize how often coronal mass ejections occur, it was probably merely an unconnected and coincidental coronal mass ejection.

Going to Extremes

One of the instruments on SOHO looks at the Sun in the spectrum between x-rays and the visible. That region, in general, is known as the ultraviolet. However, this instrument looks at the end farthest from the visible and closest to the x-rays. That part of the spectrum is often known as the extreme ultraviolet, sometimes written as XUV or EUV. The instrument is the Extreme-ultraviolet Imaging Telescope (EIT).

EIT carries a set of four filters, each of which is optimized for viewing some particular temperature region on the Sun. By comparing the four images, we can see how the temperature of the solar corona increases with height above the photosphere. Each of these filters is in the extreme ultraviolet.

A comparison of solar views through the four different filters aboard SOHO's Extreme-ultraviolet Imaging Telescope (EIT).

(NASA's Goddard Space Flight Center/EIT Science Team)

171Å
1 million °C

195Å
1.5 million °C

284Å
2-2.5 million °C

304Å
60,000 °C

The lowest temperatures are shown in the filter at 304 Å. (Remember that the lower end of visible light is at about 4,000 Å.) This filter passes primarily radiation from helium that has lost one electron. (We say that the helium is ionized or, for clarity, once ionized.) For that to happen, the temperature must be about 100,000°F (60,000°C). This temperature is about 10 times hotter than the solar photosphere but is still much less hot than the corona. It is the hottest sort of chromospheric gas. (Note that at these high temperatures, °C and kelvins are indistinguishable.)

The next lowest temperatures are shown with the filter at 171 Å. This filter passes light from eight- and nine-times ionized iron. The gas that we see through this filter is about 1,800,000°F (1,000,000°C) and is from the corona. We see particular coronal hotspots, which are located over sunspots in particular and active regions in general. We also see a diffuse background, including the bases of some of the stronger streamers.

Somewhat higher temperatures appear in EIT's third filter, at 195 Å. The gas that we see is from Fe XII and is therefore about 1,500,000°C. The hot regions of the corona are a bit more diffuse, showing how the magnetic field spreads out once it rises in the corona.

The hottest temperatures that EIT can observe appear through its fourth filter, at 284 Å. This shows Fe XV gas, which is 3,600,000°F to 4,500,000°F (2,000,000°C to 2,500,000°C). In this filter, we see clearly that the corona is bright on only some parts of the Sun and is dark on others. Those dark regions are known as *coronal holes.* They were discovered earlier with x-ray observations. The coronal holes are regions where the Sun's magnetic field is open to space instead of looping back down. Therefore, much of the gas that leaves the Sun and reaches us on Earth comes out of these coronal holes.

The coronal holes are relatively cool compared with the active corona near them. That is one reason why they are relatively dark and don't show in the filters that show the hottest gas.

> ### Fun Sun Facts
>
> A neutral atom—one with all its electrons around its nucleus—is in its basic state, which is written with a Roman numeral I. So, Fe I is neutral iron. That means that once-ionized iron is Fe II. In this way, the Roman numeral is one more than the number of electrons that have left the atom. Fe IX, therefore, has lost eight electrons.

Sun Words

Coronal holes are relatively dark regions of the corona seen in x-rays or extreme ultraviolet. They represent open regions of the coronal magnetic field, from which gas can easily escape into interplanetary space.

In these regions, the corona isn't bright, so we see down to the photosphere. But the photosphere doesn't emit continuous radiation strongly at the very short wavelengths that EIT observes, so there is no strong background of radiation.

When we discussed the observations of the visible corona at eclipses, we spoke of forbidden lines. The forbidden lines come from energy levels within the ions, from which it is relatively rare for electrons to make transitions. But in the extreme ultraviolet, we see the permitted lines, the normal lines that one would expect from gas this hot.

Fast vs. Slow

SOHO's UltraViolet Coronagraph Spectrometer studies a wide range of spectral lines in the corona. Its name shows that it not only measures the spectrum of solar photosphere, but it also can block it out, to some extent. This instrument, known as UVCS, is especially famous for having found that the solar wind comes in two types: fast and slow. The fast wind comes mainly out of coronal holes, which are often found near the Sun's poles. There, the magnetic lines of force are open to the outside. The solar wind from equatorial regions, where the magnetic field lines are looped and closed in, reaches only half that speed.

Whizzing Ions

SOHO carries a device to measure how fast ions of various atoms move in the corona. Most of the gas in the Sun—and, therefore, most of the gas in the corona—is hydrogen, so the speed of hydrogen ions is a major concern. This device also measures other ions, such as those of oxygen. An oxygen atom is more massive than a hydrogen atom, so it moves more slowly when given the same amount of energy.

Diagnosis

A good doctor tells what is ailing you in part by measuring certain vital statistics. By comparison, the Coronal Diagnostic Spectrometer (CDS) on SOHO measures vital statistics for the Sun. It takes spectra in four spectral bands that cover the ultraviolet and visible spectrum.

For example, CDS makes maps of the solar surface showing a wide variety of temperatures. In the ultraviolet, it studies neutral helium at 584 Å (20,000°C), oxygen V at 630 Å (250,000°C), magnesium IX (1,000,000°C), iron XVI (3,000,000°C), and magnesium X (1,500,000°C).

Blowing in the Wind

The expansion of the solar corona is not uniform. We speak of the outflow as the solar wind, and that wind is different in different directions.

When something is the same in all directions, it is isotropic; when something differs in different directions, it is anisotropic. So measuring differences in the solar wind in different directions is called measuring its anisotropies.

To do so, the Solar Wind Anisotropies experiment, or SWAN, observes the solar wind in different directions and, of course, over the time that it has operated. It thus maps how the solar wind varies from position to position and also over time. Like all the other instruments, it has followed changes on the Sun over the solar cycle.

SOHO Peers In

Since SOHO is in continuous sunlight, it is ideally suited for helioseismology. After all, to see a very long period, it is important to avoid nighttime! These observations involve the solar photosphere, not the corona.

Swinging Away

GOLF on SOHO isn't a game. Rather, it is an experiment that is looking for the vibrations of the Sun as a whole as part of helioseismology. Studying the Sun as a whole is known as looking at it globally. Thus, this experiment—an acronym for Global Oscillations at Low Frequencies—is looking for global oscillations.

When a wave occurs with a long period, each wavelength comes past you very seldom. Thus, its frequency is low. GOLF on SOHO specializes in these long-period waves.

Trapped

SOHO carries another helioseismology experiment, to search for oscillations at a wide variety of periods. This experiment is optimized to look for small velocity changes in which the Sun's surface rises and falls. We use the Doppler shift to see motions toward or away from us, so we mainly see these oscillations at the center of the Sun's disk.

The greatest American scientist of the nineteenth century was Albert Michelson. He measured the speed of light with high accuracy and teamed up with chemist Edward Morley to make high-precision observations of the speed of light in different directions. To everyone's surprise, it turned out that the light traveled at the same speed in all directions. This observation was the basis for the special theory of relativity that

Albert Einstein invented a few decades later (though Einstein at times denied he had known of the experiment). To do his measurements, Michelson developed the technique, known as interferometry, in which light waves "interfere" with each other. That is, as they vibrate up and down, the "up" parts can add to "down" parts of another wave and cancel each other out. One of the experiments on SOHO uses Michelson's method of interferometry to measure velocities through the Doppler shift. It is therefore called the Michelson Doppler Interferometer, or MDI.

This experiment uses its result to study helioseismology (see Color Plate 6). It is thus also named the Solar Oscillations Investigation. I don't know why this MDI/SOI name has never been simplified.

As part of its work, MDI/SOI produces images of the solar photosphere. It is also the instrument responsible for mapping the speed of rotation of the Sun under the solar surface.

Sunspots, as observed with the MDI/SOI instrument on SOHO. The white box shows a region that it studies in more detail.

(MDI/SOI Team/Stanford/ Lockheed Martin/NASA/ ESA)

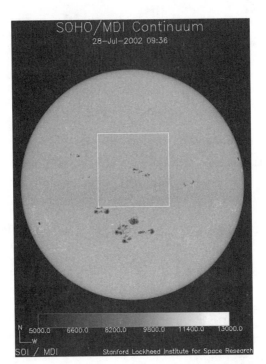

Not a Virgin

Virgo is a constellation in the sky. But the VIRGO experiment on SOHO stands for Variability of solar IRradiance and Gravity Oscillations. The study of oscillations that are controlled by the Sun's gravity is a major part of this experiment. But this experiment also measures extremely accurately how bright the Sun is. It shows how the Sun's brightness changes over time. We discuss those VIRGO results in Part 6 of the book, where we consider further the Sun-Earth connection.

ERNE but No Bert

As we have seen, the outflow of solar coronal gas known as the solar wind is a major part of solar astronomy. In Part 6 of this book, we discuss more about the solar wind's effects on Earth. Three of the experiments on SOHO are devoted to studying the solar wind.

One of the advantages of being at SOHO's position 1.5 million km (about 1 million miles) toward the Sun is that it can measure the solar wind as it passes, without the solar wind being much affected by Earth's magnetic field.

Three of the instruments on SOHO are devoted to measuring various quantities about the solar wind. These instruments are the Charge, ELement, and Isotope Analysis System (CELIAS); the COmprehensive SupraThermal and Energetic Particle analyzer (COSTEP); and the Energetic and Relativistic Nuclei and Electron experiment (ERNE). The faster a particle moves, the more energetic it is. When it gets up to be a good fraction of the speed of light, we have to use Einstein's special theory of relativity to understand its motion, and we say that its motion is relativistic.

SOHO is an example of not only international cooperation (between Europe and America), but also cosmic cooperation (between the Sun and the Earth).

A Russian-Ukrainian satellite, CORONAS-F, was launched in 2001 with several instruments that overlap SOHO's. It images the Sun in x-rays and gamma rays, measures the solar ultraviolet, and makes helioseismology observations.

The Least You Need to Know

- ◆ SOHO has been aloft since 1995, carrying a dozen solar experiments.

- ◆ Several of SOHO experiments study coronal gas.

- ◆ One of SOHO's experiments discovered hundreds of comets.

- ◆ SOHO is ideally situated for long runs useful for helioseismology.

- ◆ SOHO is in place 1.5 million km (about 10 million miles) toward the Sun to sample the solar wind.

Chapter 24

TRACEing Out the Loops

In This Chapter

◆ The Transition Region and Coronal Explorer observes fine solar details

◆ TRACE covers various temperature regimes from the photosphere on up through the transition region and corona

◆ Fine structure maps out the coronal magnetic field

◆ The solar surface is in turmoil

Nothing in the universe has a really sharp edge. There is always a place, however thin, between one thing and another. Just above the solar photosphere and the corona, the temperature arises within hundreds of kilometers from about 18,000°F to 1,800,000°F (10,000°C to 1,000,000°C). The space in which this change occurs is known as the *transition region*. Every photon of light and particle that moves out of the Sun must pass through this transition region, however thin and invisible it might be. The gas in the region emits interesting spectral lines in the ultraviolet. So when NASA sent up one of the small spacecraft in its Explorer series to study the solar corona, the spacecraft studied the transition region as well, as a bonus.

You'll Think You're Up Close

Decades ago, in the era of the series of Orbiting Solar Observatories, a decision was made at NASA and by solar space scientists to concentrate on extending the band of colors that could be observed. Spacecraft were made to study the ultraviolet and x-rays, with less concentration on showing details. Only recently did any solar spacecraft reveal details on the Sun with higher resolution than is available from the Earth.

Several attempts were made to provide a solar spacecraft with a relatively big telescope and high resolution. One of these was the 1980's Solar Optical Telescope (SOT). Unfortunately, its mirror was being made by the same contractor that made the mirror for the Hubble Space Telescope. When in 1990 Hubble's mirror was discovered to be flawed, the investigation and the taint killed SOT. A scaled-down successor, High Resolution Solar Observatory, also didn't make it to the launch pad. The coffee cup I have with its name on it is one of the few remainders of this ill-fated mission.

SOHO has a wide range of instruments and provides magnificent views of different layers of the solar atmosphere by using filters or coronagraph occulters. However, none of its instruments has really high resolution. Seeing the details of the solar corona had to wait still longer.

Sun Words

The **transition region** is the location between the chromosphere and the corona. There, solar gas rises in temperature from about 18,000°F to 1,800,000°F (10,000°C to 1,000,000°C).

Though no high-resolution solar satellite made it on to NASA's final plans for a major mission, several solar missions have become part of NASA's series of smaller spacecraft, called Explorers. One of them, named the Transition Region and Explorer Spacecraft, or TRACE, made it into space on an unmanned rocket in 1998.

The TRACE mission was planned and supervised by Dr. Alan Title of the Lockheed Martin Solar and Astrophysical Laboratory in California. Dr. Leon Golub of the Smithsonian Astrophysical Observatory, part of the Harvard-Smithsonian Center for Astrophysics in Cambridge, Massachusetts, made the high-precision mirrors. One of the problems with observing the Sun through different filters or in different ways was aligning the instruments to perfection afterward. The TRACE mission solved that problem by using only a single telescope, but having several filters coated onto the mirror surfaces. A large rotating shutter had only one quadrant open, allowing scientists to select which filter was in use. The filter selection might change, but the image size or distortions in the image wouldn't.

TRACE was sent into an unusual orbit. It orbits Earth rather than hovering in the direction of the Sun like SOHO. It is much less expensive in terms of rocket power and price to go into such an Earth orbit. But TRACE orbits from pole to pole, instead of at a slight angle to the equator, like most spacecraft. Its orbit is always perpendicular to the direction from which the Sun is shining. This *Sun-synchronous orbit* keeps the heating constant on the telescope system and keeps the Sun continuously in view without ever having a nighttime during nine months out of every year.

TRACE uses filters that are made of thin coats of metal deposited on the mirror surface. Even though its wavelengths are short, these filters and the telescope operate face-on into the beam rather than at grazing incidence. The multiple coatings not only provide filtering, but they also enhance the percentage of the extreme-ultraviolet light reflected.

Sun Words

In a **Sun-synchronous orbit,** the spacecraft goes around the Earth every two hours or so, but always in an orbit whose plane is perpendicular to the direction toward the Sun. This puts the spacecraft in continuous view of the Sun for most of each year.

Fun Sun Facts

Half an arc second on the Sun corresponds to a region about 400 km (250 miles) across, roughly the distance from Boston to New York.

The wonderful resolution of the TRACE spacecraft leads to beautiful images of these loops of gas seen in the extreme ultraviolet.

(Lockheed Martin Advanced Technology Center and NASA)

The 30-cm-diameter (12-inch-diameter) telescope was designed to provide pixels only half an arc second across, as good as or better than topical ground-based seeing. The mirror is of the Cassegrain design, in which light passes by a small mirror to hit the large main mirror. Then a reflection from the main mirror off the secondary mirror brings the solar beam back through a hole in the main mirror, where more filters and, ultimately, the detectors are located. The secondary mirror can move on a timescale of $1/10$ second, in order to compensate for jitter in the spacecraft.

Ultra and Beyond

Ultraviolet technically starts beyond the purple, at the short-wavelength end of the rainbow that people can see. But the ultraviolet light observed with TRACE is only about half those wavelengths, and it approaches the limit of what people tend to say is x-rays at 100 Å. (An angstrom—Å—is a ten-billionth of a meter and is the unit astronomers usually use for visible light and shorter wavelengths. Visible light runs from about 4,000 Å at the short end of the visible to about 6,700 Å in the red.) Here are some specifics of ultraviolet wavelengths:

- **4,000 Å:** The wavelength of the ultraviolet end of the human eye's sensitivity.

- **2,500 Å:** TRACE's filter for viewing the photosphere.

- **1,700 Å:** TRACE's filter for observing the chromosphere.

- **1,570 Å:** Another TRACE filter for observing the photosphere, passing light from neutral carbon, once-ionized iron, and continuous radiation.

- **1,550 Å:** TRACE's filter for observing the transition region by studying three-times-ionized carbon.

- **1,216 Å:** The fundamental line of hydrogen known as Lyman-alpha. It originates almost entirely in the chromosphere.

And here are some of the specifics at much shorter wavelengths known as the extreme ultraviolet (EUV):

- **284 Å:** TRACE's filter for the hottest coronal gas, that of 14-times-ionized iron, which occurs at about 3,600,000°F (2,000,000°C).

- **195 Å:** TRACE's filter for observing moderately hot coronal gas, that of 11-times-ionized iron, which occurs at about 2,700,000°F (1,500,000°C). It is also useful during solar flares, since then radiation from 23-times-ionized iron (iron that has lost all but 3 of its normal quota of 26 electrons) appears in it.

♦ **171 Å:** TRACE's filter for observing standard coronal gas, since it is sensitive to eight-times-ionized iron, which occurs at about 1,800,000°F (1,000,000°C).

The central image is a composite of the three TRACE coronal images from June 29, 1999. The surrounding images are, clockwise starting from the top: SOHO/MDI magnetic map, white light, TRACE 1,700 Å continuum, TRACE Lyman-alpha, TRACE 171 Å, TRACE 195 Å, TRACE 284 Å, and Yohkoh x-ray image.

(Lockheed Martin Advanced Technology Center and NASA)

The Sun Is Loopy

X-ray images of increasing quality over the past decades have shown that the corona seems to be made entirely of loops of gas. That is, arches of material connect two spots on the Sun. Solar hot gas is ionized, which makes it sensitive to the magnetic field. Ionized gas moves easily and freely alongside magnetic lines of force. But ionized gas does not easily cross magnetic lines of force. So the loops that we see when we look at hot gas are presumably really loops of the magnetic field. These loops resemble the shape of the magnetic field that we saw in an earlier image (refer to Chapter 2) surrounding a bar magnet.

On the Sun, there are many magnetic regions, and the regions of positive polarity and negative polarity are often close to each other. So in addition to the Sun's overall shape of a bar magnet from pole to pole, smaller regions also have this loop structure. Indeed, solar observations have found loop structures at all scales. The TRACE observations are the finest.

In order to see such fine details, the TRACE design had to limit the field of view to only part of the Sun. TRACE sees not quite a third the Sun's diameter at a time. To make a full view of the Sun at TRACE's highest resolution, a dozen images must be grouped together as a mosaic (see page 8 of the color insert).

Viewing alternately with different filters, TRACE's excellent alignment of images reveals the arrangement of loops of different temperatures. Surprisingly, different loops come down to very different conditions of magnetic field on the solar photosphere. The footpoints of these loops are sometimes in regions that are obviously of strong magnetic field, but they sometimes come down to regions where the magnetic field is not apparent.

The Sun is very dynamic, and TRACE takes image after image at a cadence of one every few seconds. When played back at 30 frames per second, the time is compressed by a factor of approximately 100, so an hour and a half of real time plays back in a minute. Even higher factors of compression can be used.

The Solar Scoop

TRACE images can be played back as movies. Samples of the movies are available to everyone on the World Wide Web. Their large-scale reproduction in the IMAX movie *SolarMax*, accompanied by artificial sound through the many loudspeakers and sub-woofers, is overwhelming and beautiful. (See vestige.lmsal.com/ TRACE/POD/TRACEpod.html.)

The movies show that the smaller loops may live only minutes, and medium-size ones may live for hours. They change rapidly. Scientists have tried to follow the heating of the loops, to determine whether they are perhaps heated at the top, with the heating then spreading down in seconds to the bottoms and footpoints. After all, if the corona is made entirely of loops, then the question of how the loops are heated is equivalent to the important question of how the corona is heated to its temperature of millions of degrees.

The answers to such questions depend not only on observations, but also on theoretical modeling. Theoretical modeling, in which a scientist these days uses not only equations but also usually a computer to help with calculations, depends on assumptions that are made to simplify the situation. The Sun is so very complicated, however, that simple situations are not realistic. A major problem is to figure out which simplifications in the model are okay to make and which affect the validity of the result.

One reason the TRACE spacecraft is such a valuable source of information is that the detail that it sees may be getting down to the actual size of the structure of the coronal elements on the Sun. Earlier images of the Sun from the ground and from space have blurred these features. Scientists have deduced average temperatures, densities, and other properties from these averages. But it may be that no actual features have conditions anything like the deduced average.

Solar Scribblings

Sometimes taking an average distorts the real situation. For example, if you have four people who make $20,000 per year and one who makes $2 million per year, then the five people together make $2,080,000 per year. The average amount of money made is the total divided by five, or $416,000 per year. Note that such an average is pretty misleading because neither the rich person nor the four poorer people have incomes anywhere near the average. The situation can be similar on the Sun when you take averages of coronal conditions.

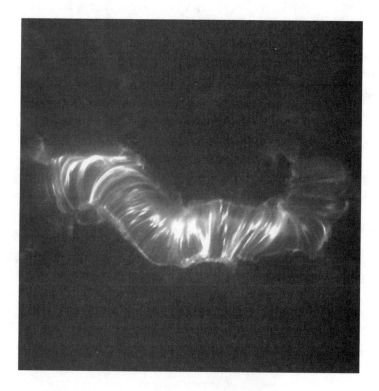

At the low resolution of the past, this object would have looked like a long filament with gas flowing along it. TRACE's high resolution reveals it to be made of small loops in the perpendicular direction. TRACE's abilities can thus prevent fundamental misunderstandings.

(Lockheed Martin Advanced Technology Center and NASA)

Fire Burn and Cauldron Bubble

Macbeth's witches may not be causing the solar coronal loops to form and move around, but the motions are quite rapid and significant. The TRACE movies have shown the solar corona to be continuously active on all scales.

In particular, the loops of gas can be seen not only to move slowly around on the Sun, but also to change abruptly. If you watch a region for a while, you may well see dramatic changes in the angles at which some of the loops appear. Since the loops merely reveal the underlying loops of magnetic lines of force, we are really seeing the Sun's magnetic field and how it changes.

The magnetic field is generated below the photosphere from a dynamo process, with moving material generating a magnetic field just as it does in a General Electric dynamo that generates power for the electric lines that bring electricity to your home. This magnetic field sticks up through the photosphere, and the coronal loops show just the tops of magnetic lines of force that are kinking up and pushing through. Solar scientists have long known that the active regions are seething with motions. Such motions have shown up in long-term movies of sunspots themselves, for example. The TRACE observations reveal the cause of the motions and turmoil in a dramatic way.

In particular, the magnetic lines of force come upward from a footpoint that has one magnetic polarity and returns downward to a footpoint that has the opposite polarity. In regions in which various loops are jumbled together, a loop may join from one footpoint to another one in order to simplify the structure. Of course, it goes from a footpoint of one polarity to another of the same polarity, since its other footpoint has the opposite polarity. Such activity is called *magnetic reconnection*.

So much magnetic reconnection is happening on the solar surface at all times that we usefully think of the surface as a *magnetic carpet*. The magnetic carpet is being studied by several instruments on SOHO, whose scientists first reported it, as well as by TRACE. Some of the energy from the magnetic reconnection comes from an electric field in the loops that accompanies the changing magnetic field. A lot of energy is released as the electric fields short-circuit.

The process of magnetic reconnection can release a lot of energy. Thus, it can be a source of the heating of the corona.

Solar Scribblings

Alan Title, head of the TRACE mission, has pointed out, "Each one of these loops carries as much energy as a large hydroelectric plant, such as the Hoover Dam, generates in about a million years!"

Sun Words

Magnetic reconnection is the abrupt change of the arrangement of loops in a region on the solar surface in order to make a lower-energy and more stable situation.

The **magnetic carpet** of the solar surface is the term that shows how common and how dynamic changes in the magnetic field are. A tangled magnetic field is found all over the images of the lower corona, and the resulting magnetic reconnection on small scales is very common.

The magnetic carpet, with the magnetic lines of force calculated and drawn in on a photograph of the solar surface. Originally, the heating was measured with the Extreme-ultraviolet Imaging Telescope on SOHO, whose image is shown, and the magnetic field was measured with SOHO's Michelson Doppler Imager.

(Lockheed Martin Advanced Technology Center and NASA)

Before it was seen to be so common, magnetic reconnection had not been expected, on theoretical grounds, to occur in the solar corona. As with so many solar phenomena, it was not predicted but had to be explained after it was found observationally. Scientists using powerful computers have now modeled the magnetic field near the solar surface in three dimensions, an advance over the simplifications introduced in the old two-dimensional calculations. The models show that convection—boiling—at the level of the photosphere brings turbulence to the lower corona. The turbulence generates electric currents that carry energy away by heating the surrounding material. Magnetic reconnection occurs during the disappearance of the currents.

The Solar Scoop

Take a virtual walk on the magnetic carpet at www.lmsal.com/magnetic.htm.

No Rolling Stones

One of the terms much discussed by some TRACE scientists is moss, a structure controversially reported in some of the images. If the old adage that "a rolling stone gathers no moss" is true, we can conclude (as we, of course, knew) that either there are no stones on the Sun or that they are rolling fast enough to leave the moss in place. Seriously, the term *moss*, as with many scientific terms, has no actual connection to its terrestrial counterpart. But on some TRACE images, the moss is seen as a low background, just as terrestrial moss on a stone may be faintly present. It appears spongy, with bright bits interspersed among darker regions. These latter regions are presumably dark because they are chromospheric gas, which doesn't emit brightly at the TRACE extreme-ultraviolet wavelengths at which the moss is best seen.

On the Sun, the moss is a thin layer of the coolest set of coronal material, which, of course, still makes it at least half a million degrees Celsius (a million degrees Fahrenheit). The moss reflects the transition region between the chromosphere and the corona. Since neither chromosphere nor corona is flat and uniform, the transition between them—and the moss—must be very structured on all levels of detail. Indeed, the moss is seen to vary in brightness from minute to minute.

The moss is seen near the bases of coronal loops. It usually lasts for a day or two in those locations. At times moss has been seen in the regions of solar flares, where it appears quickly as the postflare loops form.

Though TRACE has studied moss in detail, moss actually was seen and reported a bit earlier with a rocket-borne telescope. Ground-based images revealed the chromospheric connection, and SOHO instruments helped characterize it.

The low-level emission around the bases of the coronal loops on this TRACE image is known as moss.

(Lockheed Martin Advanced Technology Center and NASA)

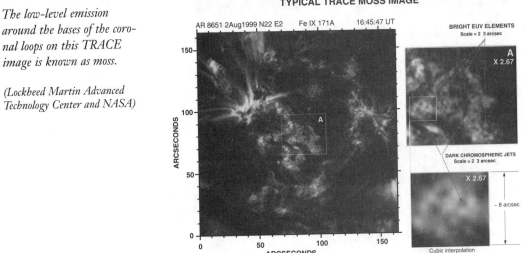

The Least You Need to Know

- ◆ TRACE, the Transition Region and Coronal Explorer, orbits the Earth to study hot gas.

- ◆ TRACE's high-resolution images reveal small-scale structure that is important to understand.

- ◆ The magnetic carpet covers the Sun with a seething set of coronal loops that are often undergoing magnetic reconnection.

- ◆ A low-level background seen with TRACE is known as moss.

Chapter 25

Plunging into the Sun

In This Chapter

- ◆ Yohkoh's successor will be even better
- ◆ STEREO will give a stereo view
- ◆ Everything's new over the Sun every day
- ◆ Probing the Sun

The images that we get hourly and daily of the Sun from a wide variety of telescopes on Earth and in space are fabulously beautiful and informative. But there is always room to do better. A series of spacecraft are planned for the next half-dozen years that should bring our understanding of solar processes to an even higher level.

B Follows A

The Japanese Space Agency had great success in its almost 10 years of operating the Yohkoh mission. The x-ray movies and flare studies at high energy gave us a continuous view of solar violence that helped us understand the underlying mechanisms. But now it has been a dozen years since Yohkoh was launched. Before launch, Yohkoh was known as Solar-A. The Japanese and their collaborators in the United States and elsewhere are hard at work on Solar-B. Only after launch will we learn its new name.

The ostensible purpose of Solar-B is to study how the Sun's magnetic field links the photosphere, where we measure it directly, and the corona. At the higher levels, we see the magnetic field only indirectly, though the form of coronal loops and streamers certainly give the shape of the field away.

The Solar Scoop

Solar-B is a joint project of the Japanese Institute of Space and Astronautical Science (ISAS), NASA in the United States, and the United Kingdom. The spacecraft itself is Japanese, and it will be launched in Japan. We hope for a launch in 2007.

Solar-B will have higher spatial resolution than previous spacecraft and will do so at a high cadence. Its field of view will encompass a sunspot region. But the main point is that its high resolution should resolve and measure the 3D magnetic field of the actual tubes of flux that are thought to permeate the corona at a finer scale than the resolution of today's telescopes. Like TRACE, the spacecraft will be in a Sun-synchronous orbit around the Earth. Over the period of its operation, we hope to understand better how the variability of the Sun's magnetic field is linked to the solar effects on Earth.

An artist's conception of Solar-B.

(NASA)

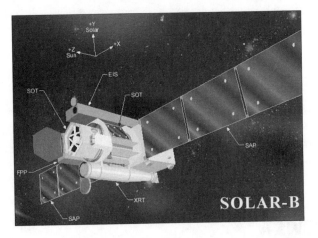

B's Visual Views

One way that Solar-B will improve on Yohkoh is to have high-resolution imaging in visible light simultaneously with its extreme-ultraviolet and x-ray observations. After all, we all like to see what is going on. Of course, there are many scientific things to see. As Yogi Berra reportedly said, "You can see a lot just by looking." And with Solar-B, we should be able to see phenomena on the Sun as small as 0.2 arc seconds (about 150 km [about 100 miles]), twice the size of each pixel. This is at least as good as the best ground-based solar telescopes, and will be available without the ground-based variations in seeing.

Leon Golub at the Harvard-Smithsonian Center for Astrophysics, who built the TRACE telescope, is building the x-ray telescope for Solar-B. Alan Title of Lockheed Martin Advanced Technology Center in California, who is in charge of the TRACE mission, is building the Solar Optical Telescope. George Doschek of the U.S. Naval Research Laboratory and Len Culhane of the Mullard Space Science Laboratory in the United Kingdom are the Principal Investigators for the ultraviolet telescope. Preparing a spacecraft like this one requires commuting a good bit between the United States or the United Kingdom and Tokyo.

The 0.5-m (20-inch)–diameter visible light telescope will provide not only imaging, but also spectroscopy. Furthermore, instruments associated with it will be able to measure the three-dimensional magnetic field at the solar surface. The magnetic field measurements will see details 10 times smaller than previous measurements of this type.

B's X-Ray Views

Solar-B will carry an x-ray telescope that is more sensitive and that has higher resolution than Yohkoh's. Its resolution of 2 arc seconds will match the best rocket images of x-rays and will be available on a minute-by-minute basis. To provide this resolution, its pixels will be half that size—1 arc second, about 2.5 times finer than Yohkoh's pixels. It will still view the whole Sun at the same time. One other improvement over Yohkoh is that it will be sensitive to the cooler, yet still million-degree coronal gas, whereas Yohkoh was optimized for the hottest gas from the corona or flares.

B's Ultraviolet

Solar-B will carry a spectrograph that works in the ultraviolet part of the spectrum. By scanning the spectrograph, the spacecraft will be able to make images of the Sun at any given ultraviolet wavelength in the covered range. It will also have a resolution of 2 arc seconds and should be able to see motions with speeds as low as 10 kilometers (6 miles) per second.

STEREO Views

We hope that Solar-B will be well situated in orbit when it comes time for NASA to launch its Solar-TErrestrial RElations Observatory, or STEREO. The mission got its name from the fact that it is actually a pair of spacecraft. Just as we humans get three-dimensional images because we have two eyes that look at an object from slightly different locations, the two STEREO spacecraft will look at the Sun from slightly different directions.

An artist's conception of the twin STEREO spacecraft viewing the Sun.

(JHUAPL)

Have you ever seen a 3D movie? Did you notice how there almost always is some time when an object appears to come quite far out of the screen toward you? The movie-makers' goal is to show off the 3D capabilities. Similarly to their methodology, the STEREO mission's pair of spacecraft are meant to observe things coming off the Sun toward us. The closer these things come to us, the more their angle will be different when viewed from the two spacecraft.

The ideal objects for such observations are coronal mass ejections, CMEs. We have said that SOHO has shown CMEs to erupt almost every day. With STEREO's sensitivity and good ability to measure distances, we should learn a lot more about CMEs and about how to predict their approach to Earth. We want to protect electrical lines, satellites, and other objects from the consequences of a massive CME hitting us, and STEREO should help us in this aim.

The general goals of STEREO are to figure out how CMEs form and what they can do to us and to the rest of interplanetary space. How do they move? How do they change over time? Coronal mass ejections are often preceded by extremely fast moving particles, which means that those particles have been given a lot of energy. How do the CMEs transfer energy to those particles?

STEREO consists of two spacecraft that will be launched on the same rocket. The spacecraft will be put in an orbit around the Sun at the same distance from the Sun as the Earth. But after a couple of months, engineers will use encounters with the Moon to put one of the spacecraft ahead of the Earth in its orbit and the other behind. Over months and years, the two spacecraft will drift farther apart. As they move farther from each other, their 3D visualization abilities will improve.

Each of the STEREO spacecraft will contain two coronagraphs for imaging the corona in visible light. Each will also be able to image the corona in front of the solar disk in the extreme ultraviolet. Additional instruments will study the space around the Sun, in part by making images and in part by directly measuring the particles that flash or drift by and the magnetic fields that the spacecraft encounter.

Activity Over Time

The Solar Dynamics Observatory, SDO, is another NASA mission that we hope for in the next half-dozen years. We hope that it will stay aloft in its Earth orbit for a good part of the solar-activity cycle. This cycle should reach its minimum in around 2005 or 2006 and its maximum in 2012 or so. It will monitor the magnetic field in particular detail, in the hope of finding the key to what triggers the changes in it that launch eruptions like flares and coronal mass ejections.

SDO's field of view will be broader than that of current spacecraft, so it will be able to image the entire visible disk in high detail at the same time. It will also often measure velocities over the entire disk as its helioseismological mission.

The spacecraft is also supposed to carry a wide variety of filters that will enable it to make high-resolution images showing all the levels of the solar atmosphere, an extension in wavelength availability and in resolution of the Extreme-ultraviolet Imaging Telescope that has been so successful on SOHO. It will also be able to take series of images at a very high cadence. SDO is scheduled to bear a coronagraph as well.

Furthermore, SDO will carry an instrument to monitor the total amount of radiation from the Sun in the extreme ultraviolet. We discuss such instruments that measure wide ranges or the totality of solar radiation in the Part 6 of this book.

Priority has been assigned, even at this early date, to getting a lot of data down from the spacecraft to Earth, given the high cadence of its very detailed imaging. It is thus planned for an Earth-synchronous orbit so that it will appear to hover over some point on Earth. To do so, it will be at an altitude of about 6½ Earth radii, the same altitude as all the Earth-synchronous satellites that bring us much of our television signals.

We hope that STEREO will be successfully sending back data from far to the sides of the Earth. So, SDO would add another observation point and help in making accurate 3D observations. At the same time, Solar-B should be providing the 3D magnetic field.

Solar Probe

To find out exactly what the Sun is doing, we want to get as close as we can. Of course, the Sun is very hot, so it is extremely difficult to get too close to it. Nonetheless, NASA has plans to do so. These plans have been canceled for financial reasons, reinstated, and then again put on hold. At this writing, I can't say whether Solar Probe will remain in the queue. It was originally supposed to be launched in 2007, travel via a gravity assist from Jupiter, and reach the Sun in 2010.

Plans are—or were—for Solar Probe to reach as close as 3 solar radii, or about 2,000 km (about 1,200 miles), above the solar photosphere. That is approximately 99 percent of the distance toward the Sun. The temperature there will be about 3,600°F (2,000°C).

By sampling the solar corona and the surrounding region as close to the Sun as possible, Solar Probe would obviously help understand the solar wind. In particular, some parts of the solar wind move away from the Sun much faster than other parts. The technical terms are the "fast wind" and the "slow wind." How does the solar corona that we see at eclipses and from spacecraft turn into the expanding solar wind? Solar Probe should also be able to see detail in the photosphere and in the corona that we just can't see from farther away. As with so many other experiments and observations of the Sun, a major goal is to figure out how the solar corona gets so hot.

A measuring instrument just can't go right up to the Sun because it would burn up or overload its sensors. The current design is to have a heavily shielded spacecraft, to keep the measuring instruments safe from the solar radiation. A mirror will move quickly out to the side of the spacecraft, in order to—one hopes—reflect a bit of bright sunlight into the instruments. Designing the survival of such a mission will be a major feat.

The Solar Scoop

Every 10 years, under the auspices of the U.S. National Academy of Sciences, a group of astronomers makes a prioritized list of the most important things to do to advance astronomy. This prioritization has proven to be an example to other sciences and has been well received by Congress and by the funding agencies. Although Solar Probe would be very expensive, the latest astronomy and astrophysics Decadal Survey has selected it as a high-priority "large" mission, given its mix of large and small missions.

Remote sensing—the technique of viewing from afar—is basically what astronomers do. But it is nice to get "ground truth" whenever we can. Comparing what we actually find in the Sun's lair with what we thought we saw from way back here on Earth should be very valuable.

The Longer Term

NASA's Sun-Earth Connection Program has a broad view of the various spacecraft above and has still more plans. Many are to investigate Earth's atmosphere and magnetosphere; we just choose not to go into them in this book. Others, more closely and directly linked to the Sun, are currently more interesting to us.

Solar Polar Imager

Even farther along the drawing board, or perhaps we should say the "dreaming board," is Solar Polar Imager. This spacecraft would go high out of the plane of Earth's orbit and look down on the poles. Only the Ulysses mission has sampled the regions above the poles so far, and that spacecraft doesn't carry any cameras.

The current plans for Solar Polar Imager is to have it orbit the Sun at about half the distance of Earth from the Sun. It will orbit the Sun three times each year—that is, its orbit will be to Earth's as 3:1. Its orbit will be very inclined, so from parts of its orbit it will be able to see the solar poles from a high angle. It is possible that Solar Polar Imager will be carried to its orbit by giant sails that pick up solar radiation rather than by the chemical rockets that we now use exclusively for spacecraft. The sails would have to be larger than a football field across, while still being very lightweight.

Though the final choice of instruments is years in the future, some ideas exist about what would be on such a spacecraft. In particular, we want to know better what the magnetic field is like near the poles and even just at a variety of relatively high latitudes on the Sun. Thus, a magnetograph is a priority for Solar Polar Imager. The spacecraft might also have instruments for helioseismology. Presumably, it would measure the velocities near the poles. Note that the Doppler effect allows us to measure only the part of the velocity of an object that is coming directly toward us or directly away from us. So up-down motions near the Sun's poles are perpendicular to our view, and we can't measure them. Solar Polar Imager would fill in that gap in our knowledge.

Not a Dodge in Your Future

Another spacecraft in the future is the Reconnection And Microscale (RAM) spacecraft. From its place in orbit around the Earth, it will take images of the solar photosphere and corona at very high resolution. The hope is that it can improve the imaging by a factor of 1,000.

The TIMED spacecraft, named for Thermosphere-Ionosphere-Mesosphere-Energetics and Dynamics. It is studying the region 60–180 km (40–110 miles) high, above the troposphere, where Earth's weather occurs, up to and including the lower ionosphere. It seeks to measure the variability of solar radiation and of Earth's response in this relatively unexplored region. TIMED was launched in 2001.

(NASA)

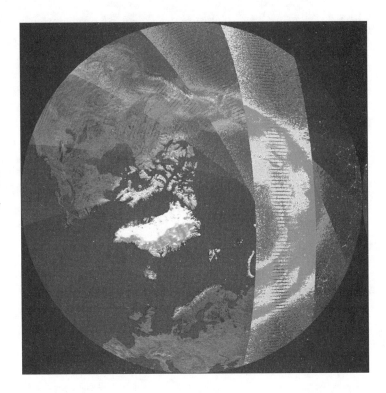

As of this writing, the hope is that RAM can image features as small as 10 km (6 miles) in the corona, compared with the 400-km (250-mile) or so lower limit of even TRACE. If it is 40 times better in each of two directions, that multiplies out to 1,600 times better overall, justifying the hoped-for factor of 1,000. It would reach 70 km (45 miles), about $\frac{1}{10}$ of an arc second, for its spectroscopy in the extreme ultraviolet. It would measure the energy in individual x-ray photons to high accuracy, with a spatial resolution of about 700 km (450 miles), or about 1 arc second.

Of course, RAM will be useful only if there really are features that small. More theoretical work is necessary to find out the odds that details will continue to be seen as we get below features a few hundred kilometers in size. Current theory indicates that this size scale will indeed be useful to study the processes of magnetic reconnection.

Solar Orbiter

The European Space Agency eventually plans to launch a Solar Orbiter. At this early date, plans are for it to orbit about 80 percent closer to the Sun than Earth and to be launched by 2010. It would be at its closest to the Sun every five months. Its orbit would be inclined so that it could observe face-on the regions reasonably far out of the Sun's equator instead of seeing them at an angle.

Because of its location out of the plane of Earth's orbit, Solar Orbiter's visible and ultraviolet-sensitive cameras would contribute greatly to 3D observations of coronal mass ejections. Of course, its location would also make it a worthy successor to Ulysses in measuring particles as they go by as well as the magnetic field at high latitude.

Solar Orbiter will carry coronagraphs to work in both visible and ultraviolet light. It will also measure the solar magnetic field with a magnetograph. Another instrument will measure the Sun's total radiation from close up. It should receive about 25 times stronger solar radiation than we get on Earth.

If tests go well on ESA's earlier SMART-1 mission to the Moon in 2003, Solar Orbiter will use the same kind of electrical propulsion rather than chemical rockets.

Can We Do It?

All space exploration is very expensive, and NASA's priorities are not necessarily those of scientists. It is impossible to predict the funding situation a half-dozen years in the future. It is clear from recent years that bringing off missions within the assigned budgets is very important. Those treasuring the prospective science returns from the missions just discussed must hope that the budgets of the space agencies of the several countries involved are robust and diverse enough to include these solar missions along with their other priorities.

The Least You Need to Know

- Solar-B, from Japan, the United States, and the United Kingdom, is the next major solar mission.

- High-resolution imaging and 3D magnetic field measurements are key.

- Plans exist to go much closer to the Sun than spacecraft in orbit around the Earth or even 1.5 million km from it and to view from high solar latitudes.

- Learning about the processes and particles that most affect Earth is a priority.

Part 6

The Sun–Earth Connection

You may think that the Sun is way out there, but really the Sun envelops us. Auroras may have been the first way we realized that the Sun and Earth are connected, but now we know of many interconnections. You may even lose your TV reception if the Sun gives out a big burp. The Sun is like a gorilla in the room, leaving us at its mercy. Through NASA's study program, called Living with a Star, we will learn more how unsteady sunlight is. And we are assessing ways in which we humans, no longer minor players on Earth, are affecting our atmosphere. The greenhouse effect and the ozone hole are things we have to think about. Furthermore, it's not a big deal to get a weather report when you leave your house in the morning. Now you can get a space weather report when you leave your planet—or even when you stay home.

26

Constancy, Thy Name Isn't the Sun

In This Chapter

- ◆ Is the solar constant constant?
- ◆ The solar constant and the sunspot cycle
- ◆ Measuring the solar constant from space
- ◆ The Sun is very round, but equatorial bright features can fool you
- ◆ The Sun and the Earth's weather

Astronomy is such an old science that it has wonderful and sometimes colorful names left over from long ago. "Sunspots," after all, just look like spots on the Sun. And they really are. "Planetary nebulae," on the other hand, don't have anything to do with planets. They are really the way the Sun and stars like it will wind up. But over 200 years ago, the first fuzzy glimpse of one looked like the fuzzy glimpses of the planet Uranus that had just been discovered. So "planetary nebulae" they were, and "planetary nebulae" they remain.

Perhaps we have a similar story with the "solar constant." Though we have long assumed that the Sun, on which we rely for our very lives, is constant, measurements made during the space age have shown us that it isn't quite so.

The Solar Parameter?

The Sun radiates astonishing amounts of energy each second. The result of nuclear fusion in its core radiates up through 71 percent of its body and then bubbles up as convection through the outer 29 percent. Once it reaches the surface, it heats the gas. That gas radiates the energy that we see, with most of the energy falling in the visible part of the spectrum.

Some of the Sun's energy comes in other parts of the spectrum than the visible. Those other parts are known to be very variable. The amount of radio flux from the Sun shows flares and other activity, and can easily be observed with radio telescopes on Earth. To study the ultraviolet, x-rays, and gamma rays, we have to observe from space, but we know that those parts of the solar spectrum also vary violently. Still, the bulk of the solar radiation is in the visible, and that appears to be pretty constant, at least during our lifetimes.

The Earth's atmosphere absorbs some of the solar radiation, and the Earth's atmosphere is variable. To find out what the Sun emits, we have to either correct our measurements for the effects of the atmosphere or go above the atmosphere. Both methods have been tried.

The solar constant is the amount of solar radiation received by each square meter of a perpendicular square facing the Sun at the top of the Earth's atmosphere, corrected to an Earth-Sun distance of 1 astronomical unit, the average distance from Earth to the Sun.

At All Altitudes

Charles Greeley Abbot, the fifth Secretary of the Astrophysical Observatory of the Smithsonian Institution, famously devoted his life to measuring the solar constant. Abbot was born in 1872 and lived to the age of 101, making measurements and publishing scientific papers right up to the end. His measurements spanned many sunspot cycles and a long period of time. Abbot started his measurements in 1895, when he was hired by the first director, Samuel Pierpont Langley, to make solar measurements. Langley, and Abbot with him, used a device—a *bolometer*—that allowed incoming radiation to heat up a cavity. They measured the temperature of the cavity electrically, since the resistivity of many metals depends sensitively on temperature. They hoped to follow changes in incoming radiation by following the electrical changes. Note that since they were basically measuring the heat from

Sun Words

Bolometer comes from the Greek word *boli*, which means "beam of light."

the Sun, their measurement included all incoming solar radiation, though, of course, the parts absorbed by the atmosphere never got through to their instruments. Still, the bolometer measured solar infrared radiation, and Langley was a pioneer in the study of the infrared.

Langley and Abbot worked hard to account for Earth's atmosphere. To eliminate as much of it as possible, Langley took a bolometer to the top of the tallest mountain in California: Mt. Whitney. Langley and Abbot also launched bolometers on balloons. Langley and then Abbot ran a Smithsonian measurement program from 1902 until 1960.

During that period, the latest re-evaluation of the Smithsonian results were that 1,353 watts fell on each square meter. In the old British units, that is equivalent to about 2 calories per square centimeter per minute. Abbot reported variations of as much as 10 percent. However, newer reevaluations of his data showed that no variations were ever really found, to an accuracy of 1 percent. Abbot had hoped to use the measurement of variations to help improve weather forecasts. He strongly believed that Earth's weather depended on the amount of energy from the Sun. That link never worked out.

> **Fun Sun Facts**
>
> Charles Greeley Abbot lived so long that people forgot that he was still alive. When Soviet scientists tried to name a crater on the back side of the Moon after him, following their first map of the far side, they ran afoul of the International Astronomical Union's rules against naming craters after living people. An exception was made.

> **Fun Sun Facts**
>
> Abbot traveled to observe total solar eclipses in North Carolina in 1900, Sumatra in 1901, a South Pacific island in 1908, and Bolivia in 1919. To assess the effect of Earth's atmosphere on the solar constant, he carried out his measurements of the solar constant from Washington, D.C., at sea level; Mt. Wilson, California, at an altitude of 1,742 m (5,715 ft.); and at Mt. Whitney, California, at an altitude of 4,418 m (14,495 ft.). His balloons went even higher.

Is the Sun Going Away?

Since the solar constant is defined as the rate at which sunlight hits a defined region at the top of Earth's atmosphere, it makes sense to go to the top of our atmosphere to measure it. Once we could get satellites into space, this type of measurement became possible.

A pioneering instrument to measure the solar constant from space was on the Solar Maximum Mission in the 1980s. To get all the solar radiation, or irradiance, the light was made to enter a cavity, where it was trapped. Because a device that measures radiation is a radiometer, the device was called the Active Cavity Radiometer Irradiance Monitor (ACRIM). The technical name for the solar constant is now Total Solar Irradiance (TSI).

ACRIM was built and controlled by Richard Willson at Caltech's Jet Propulsion Laboratory in Pasadena, and it was the first solar-constant instrument able to measure to better than a tenth of a percent. As the data came in, slight variations were seen on that level. Some big dips occurred that corresponded to big sunspots on the face of the Sun. So the longstanding question as to whether sunspots blocked enough solar energy to be detected in the overall solar irradiance was finally answered.

As the months passed, a general downward trend became visible in the solar constant. This downward trend was in addition to the day-to-day variations. From Solar Maximum Mission's launch in 1980 until it malfunctioned in 1984, the average solar constant had dropped more than a tenth of a percent. That rate of decay of incoming sunlight can't go on at that rate without severe consequences! In not many centuries, the Sun would become faint, Earth would become very cold, and our descendents would all die.

The Solar Scoop

The original ACRIM, which we can call ACRIM I, was launched on Solar Maximum Mission in 1980. ACRIM II was on the Upper Atmosphere Research Satellite (UARS), launched in 1991. ACRIM III is on a devoted ACRIMSAT mission and was launched in 1999. The satellite wasn't pointing properly at the Sun at first, but it was saved.

Fun Sun Facts

The cavity of ACRIM has mirror-like black surfaces that reflect incoming light in a way that it doesn't get out. A total of 99.99998 percent of the Sun's incoming energy in the wavelength band from 1,800 Å in the ultraviolet to 30 microns in the infrared are absorbed.

Data from ACRIM and ACRIM II showing the change in the sun's brightness over about two decades. The graph shows that, after a period from launch in 1980 through 1984 when the data weren't precise, the solar constant—the amount of energy received by a square meter of the top of the Earth's atmosphere each second—was falling. Fortunately, it began to rise in 1987, reaching a peak in 1990–91, the next maximum of the solar-activity cycle. Our hope for saving the Earth had to be in the sunspot cycle. Perhaps the solar constant was linked to it. But SMM was spinning out of control, and we had no accurate monitor. Fortunately, SMM was saved, and so was the Earth. When SMM was spun down and ACRIM started working again, the solar constant continued to decline for another year or so. In fact, the instrument was working better than ever, with increased stability of the spacecraft leading to less variation reported.

Then, in 1986 or so, we reached the minimum of the sunspot cycle. We weren't exactly holding our breaths, since it took months and years to be established completely, but the solar constant started going back up again. Finally, we knew that the solar constant merely followed the solar cycle. It goes up and down by about half a percent.

Of course, if the solar constant varies, it isn't really a constant. But, like so many astronomical terms, the term is hallowed by age and usage. We still speak of the solar constant.

 Solar Scribblings _____

Paintings from the seventeenth and eighteenth centuries show ice skating on canals in Holland, something that is now rarely possible. A "Little Ice Age" gripped at least part of Earth. Sunspots were apparently almost completely absent from the Sun for a few decades in the seventeenth and early eighteenth centuries—and people were looking for them. This period, known as the Maunder Minimum, was studied in the twentieth century especially by John A. Eddy. Is there a correlation between the absence of sunspots and the Little Ice Age?

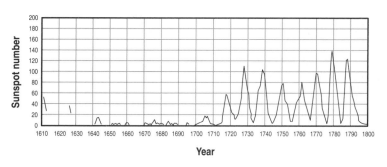

The long-term measurements of sunspots.

(J. Eddy, updated from SIDC)

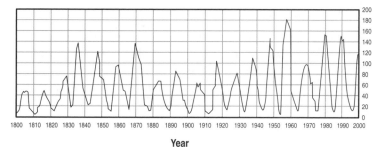

Monitoring Our Lifeline

The Sun is such an important source of energy for us that we continue to monitor it in various ways. Some other spacecraft also carried solar-constant monitors, notably the Nimbus series of weather satellites. The variations that were measured from the different satellites corresponded to each other, though there were—and remain— slight calibration problems. That is, the absolute value measured by one satellite was a steady percent or so different from the absolute value measured by another. This problem remains unresolved.

An excellent monitor in recent years has been the Variation of IRradiance and Gravity Oscillations (VIRGO) instrument aboard the Solar and Heliospheric Observatory. It carries two radiometers which, perhaps inevitably differ slightly in their results. One is the DIfferential Absolute RADiometer (DIARAD), run by the Meteorological Institute of Belgium. The other is a set of radiometers from the World Radiation Centre at Davos, Switzerland. (In German, it is the *Physikalische-Meteorologisches Observatorium Davos*, so the parent organization is known as the PMO.) They have operated since early 1986, a few months after the launch of SOHO. The measurements from the two instruments differ by 0.6 to 1.4 watts per square meter out of about 1,368 watts per square meter of solar constant. Thus, the deviation is about 0.1 percent, making the accuracy quite an improvement over the historic solar-constant measures.

NASA has independently monitored the solar constant through a series of detectors each known as an Active Cavity Radiometer Irradiance Monitor (ACRIM). We discussed an earlier ACRIM in Chapter 21 and previously in this chapter. NASA's ACRIMSAT was launched in 2000. NASA's Solar Radiation and Climate Experiment (SORCE) was launched in 2003, also to measure the solar constant very accurately. SORCE measures not only total solar irradiance (the solar constant) but also the solar output in various specific parts of the spectrum, especially as the ultraviolet, whose intensity varies more greatly than the Sun's intensity in the visible.

Occasionally, a solar-constant monitor has been flown on a space shuttle, in order to calibrate the long-term instruments that are in space. After all, the instruments on a rocket or space shuttle can be retrieved so scientists can test them to make sure that their sensitivities haven't drifted during their brief flights. They then know they can trust the calibrations made with these instruments, and compare the results with those from the instruments that have long been in space, to see if the latter's calibrations have changed.

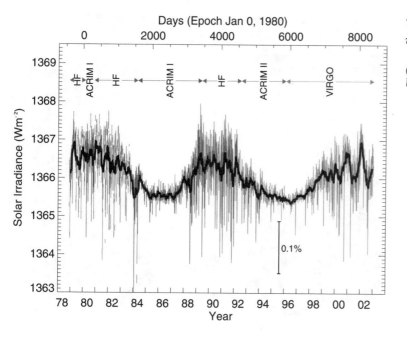

The solar constant, as it has varied, in a unified graph.

(C. Fröhlich, PMO, and the VIRGO team)

Chiaroscuro

Art historians speak of chiaroscuro, light and dark, on a painting. On the Sun, we also see a sort of chiaroscuro. There are dark regions, the obvious sunspots. But there are light regions as well. They are visible in ordinary light and stand out when you look toward the solar edge instead of toward the disk center. These light regions are called *faculae* (the plural of facula). When seen through filters passing the strong hydrogen or calcium spectral lines, the faculae within the solar limb are seen as plages. They show especially well in the calcium-line filters.

Sun Words _____

Faculae are bright regions on the solar photosphere seen in white light or through certain filters.

The light regions on the solar disk are faculae.

(NSO/AURA)

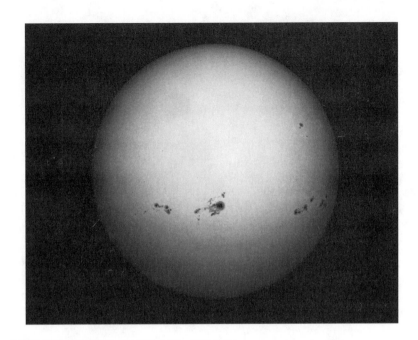

The Solar Scoop _____

Astronomers can deduce that there is magnetic activity in distant stars by monitoring their spectral lines from ionized calcium, Fraunhofer's H and K. The very central part of these lines are especially bright when there are a lot of plage areas on the stars. Long-term monitoring has shown dozens of stars that have activity cycles like the Sun's, ranging in time from a few years to longer periods.

One of the questions about the solar constant is whether sunspots cause it to drop. Similarly, do faculae cause it to rise? Does some of the energy absorbed in sunspots cause an increase in the number of faculae, compensating or more than compensating? This topic has been an important point of discussion in understanding the role of the solar constant on the Earth.

The faculae correspond to magnetic elements near the Sun's surface. They are visible to some extent in "white light"—that is, integrated visible sunlight—and also show particularly well in the light of ionized calcium. The distribution of faculae and their number change over the solar-activity cycle. The fraction of the Sun that they cover varies by a greater amount than the fraction of the Sun covered by sunspots, so faculae are very important for understanding the variations of the solar constant. Faculae and sunspots are both regions in which the magnetic field limits the amount of energy brought from below by convection. But sunspots are large enough that the area becomes cool and dark. Faculae, on the other hand, are so narrow that light comes into them from their sides, making them relatively bright.

When you look at the Sun through a hydrogen-alpha filter, you don't see the sunspots directly; you see bright active regions around them. As we mentioned briefly in Chapter 2, these active regions are known as plage, from the French word for "beach." (It is pronounced *plahje*, to rhyme with "ah.") Though the plage regions show up only in narrow wavelength bands, perhaps they have some slight effect on the total energy released by the Sun.

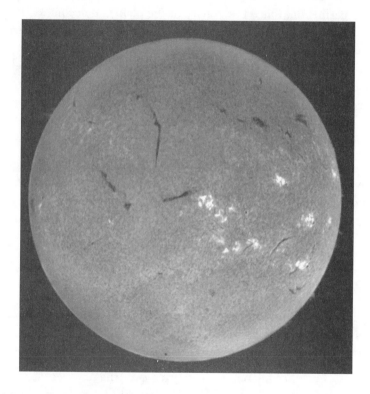

The light regions in this view of the Sun, taken through a filter that passes only the H-alpha line from hydrogen, are called plage.

(IfA/U. Hawaii)

Is the Sun Round?

The faculae affect not only the solar constant, but also the distribution of solar radiation across the face of the Sun. There are more faculae near the solar equator than there are near the poles. And the number and brightness of faculae vary over the solar-activity cycle.

Some years ago, a Princeton physicist addressed the question of whether the Sun was round—or, really, just how round the Sun is. He set up a device that blocked out most of the Sun and that rotated, with just a little bit of Sun showing around the edge. He detected a slight variation from equator to pole, concluding that the Sun was about 35 km (20 miles) larger (out of a 700,000-km 500,000-mile radius) at the equator. But how could the Sun be bulging like that? (The technical word is *oblate*.)

The physicist's theory was that the Sun had a central region that was rotating rigidly at a high speed—once every four days or so. Only that high speed in the center would account for the observed oblateness at the limb. But such a highly rotating region would cause other effects—notably, an effect on the orbit of Mercury.

The orbit of Mercury is significant because it is a test of Einstein's general theory of relativity, from 1916. The close point of Mercury to the Sun in its elliptical orbit around it is called the perihelion. Once the gravity of the planets is accounted for, the perihelion of Mercury changes by 43 seconds of arc per century. Einstein's theory accounts for that "advance," and the effect has long been considered a test of the theory. But if a rapidly rotating solar core accounted for an advance of Mercury's perihelion by 4 seconds of arc per century, then only 39 seconds would be left, no longer agreeing with Einstein's theory's prediction of 43 seconds of arc. And the physicist happened to have his own theory of gravity that was more general than Einstein's and that could be adjusted to match the effect.

Anyway, solar physicists showed that the presence of faculae near the solar limb meant that the original scientist was measuring an intensity variation but not a size variation. The rotating core just wasn't there. With hindsight, our current helioseismology measurements of the core show that the Sun does not have such a rapidly rotating core. Einstein survives.

The Sun and Weather

Does the Sun change in average brightness over many sunspot cycles, and over centuries? Such changes would have a major effect on climate. The careful measurements of the solar constant now under way should eventually settle this important question.

The Solar Scoop

As one of its major goals, NASA's Living with a Star program tries to find out how solar variability affects life on Earth. It is gathering data to answer the question of whether solar changes affect the details of weather on Earth.

Many people have tried over the past decades to link the Sun and climate. Sometimes a relationship looks good for a long time but then stops working. That effect means that the apparent relationship had been merely a coincidence. Since the 11-year solar cycle is a dominant effect, we have to wait a long time before we go through enough cycles to be sure of a relationship. We aren't even sure if the link between the Maunder minimum and the Little Ice Age of the seventeenth and early eighteenth century is a causal relation or a coincidence.

Some potential links exist between the Sun and Earth's weather. For example, the strength of galactic cosmic rays—energetic particles from beyond the solar system—that hit the top of Earth's atmosphere is linked to the solar cycle. At times of increased solar activity, the solar magnetic field deflects more of these cosmic rays away from the Earth, and the cosmic-ray flux here diminishes. Since the ionization of the atmosphere caused by cosmic rays may lead to the formation of clouds, the decrease in cosmic rays at solar maximum could diminish cloudiness worldwide, which obviously affects weather. Most directly, less cloudiness means more sunlight penetrating to the Earth's surface.

Furthermore, the amount of ultraviolet, which affects things like the ozone layer (which we discuss in the following chapter), is linked to the solar cycle. Some of the spacecraft that monitor the solar constant also carefully monitor the ultraviolet flux that arrives at Earth from the Sun for that and other reasons. In particular, the ultraviolet light from the Sun affects the rate at which ozone is formed and, at different ultraviolet wavelengths, the rate at which it is destroyed. The absorption of ultraviolet radiation by ozone also affects the temperature in the higher levels of our atmosphere. But nobody has ever linked the Sun's radiation, and variations in it, directly to the daily weather or, more particularly, to the weather at particular locations.

The Least You Need to Know

- The solar constant is the amount of energy received by each square meter of the top of Earth's atmosphere each second.

- The solar constant was found to vary by a fraction of a percent through space monitoring.

- The solar constant varies slightly through the solar-activity cycle.

- The solar constant can have a long-term effect on Earth's climate, but no effect on daily weather is known.

- Cosmic rays are affected by the solar-activity cycle and may affect the rate of cloud formation and thus weather.

- The amount of ultraviolet radiation varies by a much greater factor than does the solar constant and may affect weather.

Chapter 27

Greenhouses of Salt

In This Chapter

- ◆ Solar energy warms Earth beneficently
- ◆ The atmosphere balances energy coming in with energy flowing out
- ◆ The greenhouse effect is increasing because of human contributions
- ◆ The ozone hole over Antarctica admits solar ultraviolet at certain times of year
- ◆ Controlling the release of Freons and certain other gases should heal the ozone hole

We may get an occasional cold wave in many parts of the United States, with temperatures of 0°F (–18°C), but on the whole, the temperature is within a comfortable range. Earth is warmed by energy from the Sun. But a steady contribution from trapped radiation—the "greenhouse effect"—also makes life on Earth possible for us. Many people are now worried that too much of a good thing may be upon us. Also, an entirely separate effect, the ozone hole, opens over Antarctica each Antarctic springtime and sometimes extends farther north.

Energy In and Out

Most things in the universe are in a state of balance. Without this condition of equilibrium, everything would be changing all around us. For example, the Sun itself is in balance, with gravity pulling inward on its gas at the same strength as pressure from its internal energy pushes out.

In the Earth's atmosphere, we have a balance. Energy from the Sun comes in, and energy from the Earth goes out. If the energy balance is not exact, then the Earth's atmosphere would heat up or would cool down. We know we are more or less in balance each year since we don't see a major trend in temperature. But serious investigation seems to be showing that there is, in fact, a slight trend in temperature. We may not notice it year to year, given the natural variation in temperature and climate, but over decades or a century, it looks as though it will amount to quite a significant effect.

Planetary scientists can calculate what the temperature of the Earth should be, given the value of the solar constant and the Earth's distance from the Sun. They merely compare the energy coming in with the energy going out. The energy coming in is the solar constant times the area of the Earth's disk. The energy going out follows the same black-body law of temperature that the Sun itself follows. The hotter an object is, the more total energy it gives out.

We can consider the energy balance of Earth. Let us first imagine that the Earth had no atmosphere. Then Earth's surface would merely heat up until it radiated the same amount of energy coming in. It is easy for physicists to calculate that temperature. The planet would be very cold—too cold for us to be comfortable on Earth.

But we can measure the Earth's average temperature, and it comes out about 60°F (33°C) warmer than the temperature we would have without an atmosphere. We are warmed by the presence of the atmosphere to the livable Earth climate that we have.

The Terrestrial Greenhouse

How does the atmosphere do this warming? It does so by trapping the solar radiation. But we know that the atmosphere is transparent to incoming sunlight. The trick is that Earth itself transforms the sunlight from the incoming wavelengths to wavelengths that don't pass through the atmosphere.

Let us return to our toaster, which we used as an example in Chapter 8. When you turn it on, it glows faintly and then eventually glows red hot. Objects that are being heated from cold temperatures start radiating most of their energy in the infrared. As they are heated, more of the energy is given off at shorter infrared wavelengths. Some even appears in the longer part of the visible spectrum, the red. Eventually, if we heated an object to the 11,000°F (6,000°C) or so temperature of the Sun, most of the energy would appear in the visible, centered in the yellow and green part of the spectrum.

Greenhouse Gases

But for terrestrial temperatures, most of the energy is in the infrared. The sunlight comes through Earth's atmosphere, which is transparent at the yellow and green wavelengths where the Sun gives off most of its energy. Then the Earth's surface heats up and radiates mostly in the infrared. Something in the Earth's atmosphere keeps that infrared radiation from getting through. Eventually, the atmosphere heats up sufficiently that just enough infrared energy gets out to balance the incoming visible radiation. And that is where our atmosphere sits.

What traps so much of the infrared radiation? It has to be some gas or gases in our atmosphere. The gas that captures most of the outgoing radiation is water vapor. And infrared radiation is also strongly absorbed by the gas carbon dioxide. The carbon dioxide absorption fills the gaps in the spectrum of water vapor through which infrared would otherwise get out. Between the two gases, little infrared escapes. Increasing the "greenhouse gases," especially carbon dioxide, warms the atmosphere enough so it holds more water vapor.

The spectrum of carbon dioxide shows some strong spectral lines in the visible. In fact, for molecules, the absorption is spread over a range of wavelengths rather than being at narrower bands of wavelengths, as for ordinary atoms. We thus speak of molecular bands, the wide ranges of wavelengths that molecules absorb. Because the molecular bands are broad, infrared energy from the warm Earth doesn't get out.

> **Fun Sun Facts**
>
> The Earth's atmosphere is about 75 percent nitrogen molecules and about 20 percent oxygen molecules. It is the oxygen that we breathe. The rest includes carbon dioxide and various other gases, to lesser percentages.

Not Your Ordinary Greenhouse

How do we describe the trapping effect of the Earth's atmosphere for infrared radiation emitted by the Earth's surface? People make an inaccurate analogy to a greenhouse on Earth, where the air inside the greenhouse is warmer than the air outside. It is easy to think that a terrestrial greenhouse is warm because its glass absorbs infrared. If that were so, we could say that solar energy passes through the glass—which is transparent in the visible—that it heats up the ground, tables, plants, and whatever else is inside until they radiate in the infrared, and that then the infrared is trapped. (Even so, eventually, the energy getting out must balance the energy coming in, and the temperature stabilizes at a new, higher, level.) Unfortunately for this easy analogy, that isn't the major effect for a terrestrial greenhouse. Greenhouses on Earth get warmer inside primarily because the surface keeps outside air, which we can think of as wind, from mixing the warm air that is formed inside with cooler outside air.

The percentage of carbon dioxide in our Earth's atmosphere has doubled in the last hundred years or so and continues to rise year by year. (The graph shows a yearly cycle, but the trend is clearly strongly upward.) These observations continue to be made at the Mauna Loa Observatory in Hawaii.

(Pieter Tans, NOAA)

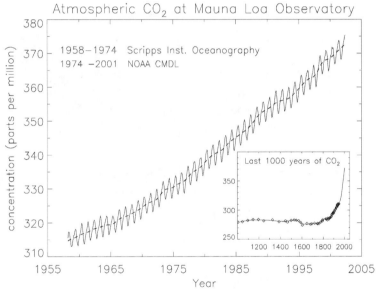

Whether or not the effect is the same in the Earth's atmosphere as it is for an actual greenhouse, the trapping effect is known as the *greenhouse effect*. The greenhouse effect works on other planets as well as on the Earth.

The importance of the greenhouse effect was realized in part by studying the atmospheres of the planets Venus and Mars. Studying them helped show that our models for how planetary atmospheres were affected by various changes were correct. For example, substances suspended in the air—sulfates for Venus and dust for Mars—have effects on global warming.

The Solar Scoop

A team of scientists with which I work has found signs of global warming on both Neptune's moon Triton and Pluto. We watched Triton and Pluto pass in front of stars and carefully measured the rate at which their atmospheres dimmed the starlight.

Sun Words

The **greenhouse effect** occurs when incoming sunlight passes through a transparent atmosphere and is transformed to infrared radiation that is partly blocked from escaping by the atmosphere. Global warming is the upward trend in temperature. Most scientists have concluded that most of it comes from the greenhouse effect caused by human contributions of carbon dioxide and other gases to the Earth's atmosphere.

A runaway greenhouse effect, like that on Venus, occurs when the blocking in the infrared is so severe that the atmosphere heats up drastically.

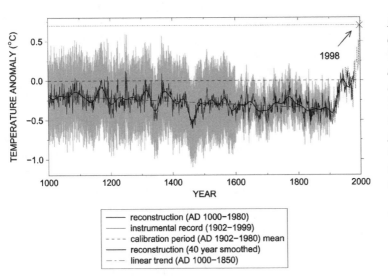

legend:
— reconstruction (AD 1000–1980)
— instrumental record (1902–1999)
- - - calibration period (AD 1902–1980) mean
— reconstruction (40 year smoothed)
- - - linear trend (AD 1000–1850)

Earth's average temperature over the last millennium. The rise in the last century appears to be too much to come from natural causes. The scientific consensus is that much of it is from the greenhouse effect caused by carbon dioxide introduced into our atmosphere by us humans.

(Michael E. Mann, DES, U. Virginia)

Venus's atmosphere is 96 percent carbon dioxide, and its greenhouse effect is therefore much more significant than Earth's. The temperature on Venus reaches about 900°F (about 500°C), hot enough to melt lead. It is a hellish atmosphere now, perhaps with rain of sulfuric-acid droplets to make it even worse than it would be with merely the high temperature. Nobody has suggested that our Earth's greenhouse effect would run away the way Venus's did. That's a good thing, because though Venus and Earth are near twins in terms of size, we don't want to

Solar Scribblings

Billions of years ago, Venus's atmosphere was probably very nice. Perhaps life even evolved there. Though most efforts have been placed on searching for life on Mars, or even on Jupiter's moon Europa, a few people wonder whether life arose long ago on Venus and is surviving deep underground or in clouds.

become like Venus. Studying the atmospheres of Venus and Mars helps us understand our own atmosphere. Though nothing so totally hellish as Venus is likely, global warming could make extra greenhouse gases slowly evaporate from frozen arctic soil and seabed muck, bringing yet more warming and vast harm.

We Are Changing Our Atmosphere

Since the beginning of the Industrial Revolution on Earth about 200 years ago, we humans have been sending greenhouse gases into our atmosphere. The largest contribution is the gas carbon dioxide, the byproduct of many burning processes and engines. A secondary greenhouse gas is methane. The absorption bands in its spectrum tend to plug up some of the gaps in the absorption bands in the spectrum of carbon dioxide. The windows of transparency in the infrared between absorption bands of carbon dioxide are closed by the absorption by methane.

> ### Solar Scribblings
>
> Many sources of greenhouse gases are natural. Carbon dioxide and methane are given off by decaying plants, for example. Cows give off methane in their flatulence, after digesting their food. So we can't cut off all sources of greenhouse gases, but we can limit the amount that humans contribute.

> ### The Solar Scoop
>
> The Earth may have gotten most of its water from comets. Billions of years ago, the rate of comets hitting the Earth was a lot higher than it is now, and they carried tons of water.

A major source of greenhouse gases now comes from fossil fuels. These fuels—chiefly oil, gas, and coal—are today's remnants of plants and animals that died millions of years ago. The remains were transformed into oil, gas, and coal, which we can now mine. But using these fossil fuels releases carbon dioxide into the atmosphere. The amount of carbon dioxide in the atmosphere has increased greatly over the past couple of centuries.

Finding two things that both increase and decrease at the same time doesn't necessarily mean that one causes the other. But there are persuasive links between the amount of carbon dioxide in the atmosphere and the Earth's overall temperature. Still, there are other possible contributions. Changes in the solar constant over time are something to consider. The measurements of the solar constant find a variation of about 0.1 percent over the sunspot cycle, but perhaps there has also been a variation of some tenths of a percent over centuries.

Can the increase that has been measured in the Earth's average temperature be attributed to changes in the solar constant rather than the greenhouse effect? Though a handful of scientists disagree, the overwhelming consensus is that the link between the greenhouse effect and the Earth's average temperature is definite. Careful consideration indicates that much of the gain in the Earth's average temperature measured in the past half century comes from carbon dioxide and other gases that humans have injected into the Earth's atmosphere, largely through burning fossil fuels.

While it is hard to predict the future precisely, the consensus, based mainly on super-computer studies, is that the Earth's average temperature is likely to rise several degrees by the the end of this century. Of course, that rise won't be uniform; some places will become a lot warmer, others will cool, and others will stay the same.

The consequences that we foresee of increasing temperature can be severe. It isn't just that growing seasons are made longer or bands where certain crops or other plants grow are pushed toward the poles. Severe storms—hurricanes and tornadoes—may become more common. Melting of ice caps in the Arctic or Antarctic and warming of the oceans will lead to a rise in ocean level. Whole island nations and coastal regions of other nations will be submerged. And the effects can be very widespread. Little realized is that New York is on a level with southern Europe, on the same latitude as Madrid. Cities like Paris and London are much farther north than New York, yet they have similar climates. The northern European cities are warmed by winds crossing the Atlantic Ocean, where warm surface water moves north from the tropics (the Gulf Stream), then cools and sinks near Iceland. Many scientists think that global warming could halt this circulation, bringing a big chill to lands all around the North Atlantic.

Sometimes a few particular examples drive home a point better than even strong general warnings. A photograph of open water at the North Pole did so. It turned out that pack ice drifts about, and it has long been common to find open water at that cardinal point; however, it is true that the thickness of the northern polar ice cap has diminished by a factor of 2 in a decade. Glaciers in Alaska and the Alps and elsewhere are retreating and may soon disappear. Even Mt. Kilimanjaro, the tall mountain standing on the equator in Africa, famous for its ice cap in Hemingway's *The Snows of Kilimanjaro*, may soon be bare.

What Can We Do?

How can we limit our human contribution to greenhouse warming? International meetings are devoted to the topic. There is widespread consensus that our use of fossil fuels must be limited. But the economic consequences of such limiting can be debated, and political questions arise over just how much we should do now. And can we compensate for greenhouse gases by planting trees, since trees take in carbon dioxide and give out oxygen? The emissions and absorptions by trees are undergoing increased scrutiny. Cutting down forests, for example, can leave a lot of material on the ground to decay, releasing its carbon dioxide into the atmosphere.

Africa's Mount Kilimanjaro. Its perennial ice cap may melt in a decade or so, as a result of global warming.

(Jay M. Pasachoff)

The Solar Scoop

A 1997 treaty was negotiated in Kyoto, Japan, specifying how nations should cut emissions of carbon dioxide and other greenhouse gases. Several international meetings have been held since on how to carry out this Kyoto Protocol. But the matter has become very politicized. The United States, in particular, has refused to ratify the Protocol. One issue raised especially by the United States is whether developed countries should be held to the same standards as the developing world.

Energy sources that do not contribute greenhouse gases should be adopted. Energy directly from the Sun causes the wind to blow and the tides to come in and out. Wind farms and tidal stations are being developed all over the world. But these currently are on a much smaller scale than the many fossil fuel power plants.

Direct usage of solar energy is increasing. Solar cells exist that take in solar energy and put out electricity. But they currently are so expensive to make that, even though they cost nothing to use, the total cost of energy using solar cells is prohibitive except for isolated locations where it is expensive to install power lines.

Nuclear energy, which provides over 15 percent of Earth's energy needs, is the sole major method of providing a lot of energy without giving off greenhouse gases. Many scientists think that nuclear energy usage should be increased for this reason and that the objections to the storage of nuclear waste products are overblown. But these objections are widespread, and there is a significant "nuclear fear," to use a psychological term popularized and explained by historian of science Spencer Weart. Nuclear proliferation issues, with worries that nuclear reactors could provide material for nuclear bombs, are also important. And there are fears of exploding reactors, with the Chernobyl runaway historically being the major nuclear debacle.

The holy grail for many people is nuclear fusion, the process that fuels the Sun and the stars. In fusion, hydrogen itself is the fuel, and the oceans are full of it as part of each molecule of water: H_2O. Some major international research projects are underway to develop fusion reactors. But what the Sun does easily 93 million miles from us is hard to reproduce on Earth. There are those who say, "Fusion is 50 years in the future, and always will be." Let us hope that this cynical comment is wrong.

Hole in the Sky

The people of Puenta Arenas, Chile, bundle up when they go out. But they bundle up whenever they are in the Sun, especially in their springtime. They slather themselves with highly absorbent suntan lotions. (Spring, at their southern-hemisphere location, is in September and October.) They are concerned less with the cold than with the ultraviolet radiation that is coming from the Sun. For they are in one of the prime cities troubled with what has come to be called the *ozone hole*.

Ozone is a simple gas, a form of oxygen. Whereas the oxygen that we breathe is the O_2 oxygen molecule, ozone is O_3. This ozone molecule has both favorable and unfavorable consequences for us on Earth. At ground level, the ozone is a pollutant that we find, for example, in automobile exhaust. It isn't good to breathe. But high in the atmosphere, at an altitude of about 25 miles, a layer of ozone protects us from the Sun. This ozone layer absorbs the ultraviolet light that would otherwise come through to hit us, increasing skin cancer and other problems.

Sun Words

The **ozone hole** is a thinning of the ozone seen in the Antarctic region each southern-hemisphere spring. It results from the breakdown of ozone molecules by chlorofluorocarbons, or CFCs.

Scientists Sherwood Rowland, Mario Molina, and Paul Crutzen received the 1995 Nobel Prize in Chemistry for their work on the ozone hole. Crutzen did the basic chemistry about how ozone is formed and decomposes in the atmosphere. Rowland and Molena noted the threat to the ozone layer based on Freons in the atmosphere. They published an article about it in 1974, and some limitations on Freons resulted. A major shock then came in 1985, when Joseph Farman and colleagues in England realized that measurements over Earth's South Pole started showing a drop in ozone content around 1980. The ozone level dropped year by year in an increasingly wide area. The region of low ozone came to be called the ozone hole, though it is a hole in only a figurative sense—similar to our astronomical use of the term *window of transparency*, referring to a wavelength band.

Fun Sun Facts

Ultraviolet is the name given to wavelengths shorter than about 4,000 Å, just short of the blue, and extending down to x-rays at about 100 Å. The ultraviolet between about 3,000 Å and 4,000 Å comes through our atmosphere, although we can't see it with our eyes, except for a glimpse of the extreme long-wavelength end. That radiation gives us our suntans. UV-A is from 4,000 Å down to 3,200 Å. UV-B is from about 3,200 Å down to 2,900 Å. When you buy sunglasses, look for pairs that are certified to block 99 percent of both UV-A and UV-B, to impede the passage of ultraviolet that can eventually lead to cataracts. The ozone completely blocks over 90 percent of the UV-B as well as even shorter wavelengths of ultraviolet, those shorter than 2,900 Å, known as UV-C.

NASA has an ambitious program of monitoring ozone. Several satellites take measurements of ozone on a daily basis, and the results are available for all to see on the web. Analysis of the process has shown that certain molecules rise into the stratosphere and that they attack ozone molecules, breaking them apart. The chief set of culprits are molecules known as chlorofluorocarbons. These are molecules that contain chlorine ("chloro-"), fluorine, and carbon atoms. (Freon is a common trade name for them.) The molecules were discovered in the 1930s and were thought to be inert—that is, unchanging. Therefore, they were considered harmless when they were used for many purposes, especially refrigeration and air conditioning. The discovery in the 1970s that they broke apart stratospheric ozone came as a shock.

Other gases also affect the ozone layer when they get up there. Halons are a serious problem as well. Some fire-fighting systems—such as those in libraries, where flooding with water would be disastrous—use these halons, which are compounds based on bromine rather than on fluorine.

We now know that the CFCs and halons aren't entirely inert. They break down when they get into the stratosphere. Then the chlorine and bromine atoms interact with the ozone molecules, breaking them down into oxygen, chlorine-oxygen molecules, and so on. These new molecules do not have ultraviolet-absorbing power.

The ozone hole is the result of what happens during the long, cold Antarctic winter. Ice crystals form in the air, and the molecules that arose from breaking down CFCs and halons attach themselves to the ice crystals. When sunlight hits these cold crystals each Antarctic spring, as the months-long nighttime ends, it starts chemical processes that break down the ozone very efficiently. Atoms such as chlorine, which break down the ozone, are left over after each reaction to break down still more ozone. Both sunlight (and its variation over the seasons) and the interacting molecules and atoms are needed to make this effect. The more CFCs and halons there are in the atmosphere, the deeper the ozone hole can become and the farther north of Antarctica it can extend.

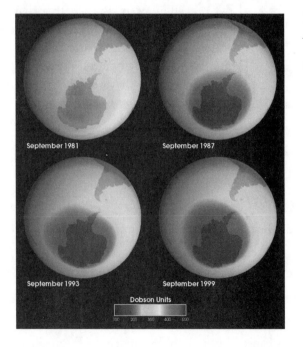

The growth of the spring Antarctic ozone hole.

(NASA's GSFC)

The ozone hole has been restricted to southern regions—at least, so far. Though some drop in atmospheric ozone has been detected over northern polar regions in northern hemisphere spring, the effect has never been as deep as it has been in the southern hemisphere. Most scientists, therefore, have not given the name "ozone hole" to the northern effect.

Many steps have been taken to control the release of CFCs. For example, they were often used in propellants in cans of things like shaving cream. They have been replaced by compressed air or by other gases for that purpose. Newer air conditioners substitute other molecules for CFCs. These substitute molecules are apparently not as efficient as refrigerants, but they are safer for the environment. The use of CFCs in making computer chips is being better controlled as well.

Steps also are being taken to keep CFCs already in use from making their way into the upper atmosphere. For example, recycling is required for the CFCs in car air conditioners. Use of CFCs in household air conditioners and refrigerators is restricted, though use of recycled CFCs as opposed to newly made ones can continue until 2020.

Fun Sun Facts

Ozone in the atmosphere is measured in Dobson units. G. M. B. Dobson was a British atmospheric physicist. Each Dobson unit gives the amount of ozone in a column that is 1 cm² in size and that extends upward through the entire atmosphere.

The ozone hole got bigger year by year through 2001. Strangely, the 2002 hole did not continue that sequence. It split in two parts and never got as deep as preceding years. But whether that is an aberration (theories for why it was special, based on an unusual stratospheric weather pattern, have been advanced) or the beginning of a trend remains to be seen. Predictions are that with the current steps taken to control the release of ozone, Earth's ozone layer should be healed by about 2050. That would be a success story for the reclamation of part of our damaged environment as the result of scientific vigilance.

The Least You Need to Know

◆ The greenhouse effect is the natural warming of the atmosphere by trapping sunlight.

◆ The greenhouse effect is life-giving on Earth but has run away on Venus.

◆ Human-caused contributions of carbon dioxide have increased the greenhouse effect.

◆ The ozone hole is caused by sunlight hitting chlorofluorocarbon molecules in the cold Antarctic air each spring.

◆ Steps taken to limit ozone-depleting gases, such as Freons, should heal the ozone hole.

28

The Forecast Today Is Flares

In This Chapter

- The Sun flares up, linked with the sunspot cycle
- Solar flares are followed with telescopes on the ground and in space
- The corona burps regularly
- Solar telescopes often pick up comets near the Sun.
- Space weather affects the Earth

How's the temperature today? Will it rain? Will a rain of particles from the Sun make your radio go haywire? Should you bring your airplane down to lower altitude, or bring your astronauts back down quickly? In this twenty-first century, we must forecast not only weather near the Earth's surface, but also space weather.

How's the Space Up There?

Buck Rogers, Captain Video, and other early fictional astronauts moved easily around the solar system. But the solar system, outside the Earth's protective envelope, is a rough place. High-energy particles of light, including x-rays and gamma rays, and fast-moving particles of matter smack into whatever is up there. To function in space, we must be alert to *space weather*.

Sun Words

Space weather is the condition of space in the vicinity of the Earth as it is affected by particles and radiation from the Sun.

Sun Words

A solar flare is an abrupt brightening of part of the Sun within seconds, including the release of electromagnetic radiation and particles.

Our knowledge of space weather dates back to 1859. Then, the English amateur astronomer Richard Carrington noticed that a bit of the Sun's surface brightened abruptly. He was, in this way, the first person to see a *solar flare*. Carrington was observing in ordinary sunlight, with a projected image that cut down the overall intensity over direct observation. What he saw was a white light flare, and he reported that a large magnetic storm followed this flare. Scientists today still observe flares, but we mainly study them in other parts of the spectrum. They have long been monitored in hydrogen light in the visible part of the spectrum, and radio flares have also long been under study. In recent years, satellites have provided monitoring of solar flares in x-rays and gamma rays. Solar flares in white light are very rare; only about 50 have ever been seen.

Richard Carrington's 1859 observation of the first solar flare to be seen. He saw a brightening at points A and B that soon moved to points D and C.

(MNRAS)

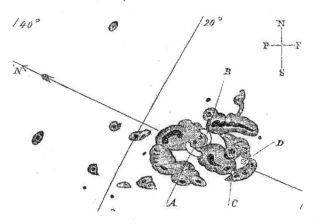

Monitoring the Sun in hydrogen light, as has long been carried out at solar observatories, has provided movies of solar flares. In views looking downward at the solar disk, the brightening appears abruptly and spreads along magnetic field lines. The first brightening takes place in seconds, and the whole region may then remain bright for hours. When seen on the solar limb, flares explode outward into space. While some of its matter falls back, much of the matter is ejected from the Sun at high speed. The coronal mass ejections discussed in Chapter 23 used to be thought of as caused by flares, but the present view is that the ejection and the flare are both aspects of a large-scale rearrangement of the magnetic field in the corona.

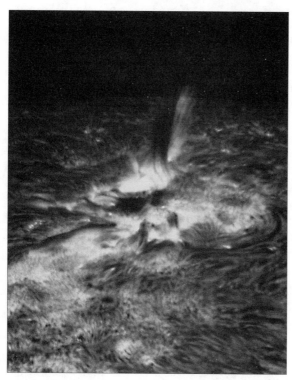

A flare seen in hydrogen-alpha, the red line of hydrogen, from the Big Bear Solar Observatory. The material twisted as it rose, following the magnetic lines of force.

(BBSO/then Caltech/now NJIT)

Twisted Magnetism

A single solar flare can brighten the solar output by as much as 0.1 percent within seconds. How does so much energy accumulate, and why is it released? More technically, astronomers speak of two things:

◆ The energy-storage mechanism

◆ The trigger

The frequency of flares varies along with the solar-activity cycle. Many more big flares occur at sunspot maximum.

Though the details of explaining flare activity are controversial, it is clear that the energy for a flare is stored in the Sun's magnetic field. The powerful magnetic field that is typical in sunspot regions keeps the matter restrained and allows the energy to build up. Then magnetic reconnection occurs. That is, the positive and negative polarities seek a new configuration with respect to each other. As they become connected in a different series, energy is abruptly generated and heats the corona to tens of millions of degrees.

We have already seen how scientists using the SOHO spacecraft discovered a magnetic carpet on the Sun of readily changing magnetic connections. The details of the flare changes can best be followed spatially with the TRACE spacecraft. TRACE observations show how the flaring begins at a particular location and then spreads over a wider region. The high resolution of TRACE, in particular, also shows how thin post-flare loops form.

The evolution of a flare as seen from TRACE, including the formation of postflare loops.

(Leon Golub, SAO, and Alan Title, LMSAL)

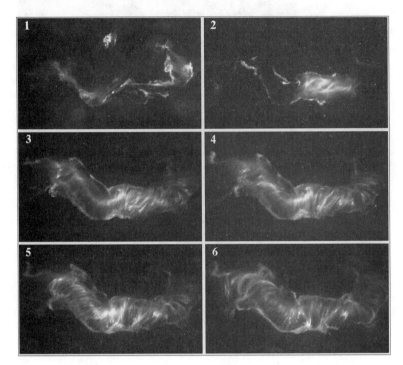

Particles ejected from the Sun in solar flares move so fast that they can reach Earth in hours. But x-rays and gamma rays formed as the flare begins travel even faster, since they move at the speed of light. They pass the 150,000,000 km (93,000,000 miles) in only eight minutes. On some spacecraft, such as Yohkoh, when an x-ray camera sensed a flare, it started other instruments recording at a higher cadence than normal.

X-ray observations are so important to flare studies that the current classification of flares depends on them.

- ◆ **X-class:** The flares with the strongest x-ray emission are X-class. They can lead to major geomagnetic storms and to massive disruptions of communications on Earth, especially when associated with coronal mass ejection directed toward the Earth.

◆ **M-class:** The medium-size flares, M-class, are less significant in general, though some particles from the Sun can arrive at the Earth, and minor geomagnetic storms or radio blackouts can occur.

◆ **C-class:** The smallest type of flares, C-class, doesn't affect us much on Earth.

These types are classified by the strength of emission in the range from 1 to 8 angstroms.

Flares are also classified by the area that they take up in H-alpha, from an importance of 4 for the largest flares in area down to an importance of 0. Optical flares are also classified by B for brilliant, N for normal, and F for faint. A huge flare, X-class in x-rays, might also be of importance 3B based on its optical appearance.

Flares are also monitored in the radio part of the spectrum. The radio comes through a window of transparency, so it can always be observed from Earth's surface. Links of radio telescopes like that at the Nobeyama Radio Observatory make images of the flares on the Sun. But a single flare makes so much radio energy compared with the radio background of the quiet Sun that flares can be monitored even without any resolution of details on the solar surface.

Flares and Other Eruptions

When you look at the limb of the Sun in the light of hydrogen-alpha, even without an eclipse, you usually see small regions sticking out. These are almost always prominences, gas at about the temperature of the solar chromosphere, approximately 20,000°F (10,000°C). Prominences look bright because we see them silhouetted against the dark sky. Likewise, when we look down on prominences, seeing them silhouetted against the solar photosphere, they look dark. These disk features are called filaments and run along the neutral line that separates opposite magnetic fields.

Prominences (and, therefore, filaments) can live for months. The longest lived of these quiescent prominences can be seen repeatedly each time the Sun rotates enough to bring them to the limb. But sometimes a prominence erupts. These eruptive prominences lift off the Sun, usually gently, and can extend millions of kilometers.

Just because you see something at the edge of the Sun, don't call it a flare. Too many people talk of the "flares on the Sun" when they really mean the prominences. Flares are much more explosive and powerful events than erupting prominences. Temperatures in flares reach millions and even tens of millions of degrees; they are much more impulsive than erupting prominences.

A large quiescent prominence at the limb of the Sun. Prominences can be seen with the eye as totality begins and ends at solar eclipses, or through hydrogen-alpha filters outside of totality.

(BBSO/Caltech, now NJIT)

Ejecting Matter

From time to time over the past 150 years, the corona at eclipses has seemed oddly asymmetric. Features appeared that didn't seem to extend radially outward. For a long time, these features were thought to be merely unusual and odd. Only since the launch of Solar Maximum Mission in 1980 was it realized that these coronal mass ejections (CMEs) are fundamental parts of the solar activity. Before this time, the ground-based coronagraphs didn't see out far enough above the Sun to realize that the ejections occasionally seen were so major or common.

Coronal mass ejections represent part of the corona drifting off into interplanetary space. They are carried by gently erupting magnetic field. Each coronal mass ejection carries perhaps 10 billion tons of matter into space at a speed that averages 400 km (250 miles) per second. The matter can reach Earth in days. The fact that the speed of the coronal mass ejection can increase as it gets farther from the Sun shows that it is controlled by the coronal magnetic field.

The Large Angle Spectroscopic Coronagraph (LASCO) set of telescopes on SOHO has proven especially valuable for studying coronal mass ejections. The telescopes show that at peak times of the solar-activity cycle, coronal mass ejections go off about every day.

LASCO on SOHO has discovered over 500 comets, which often appear in the same field of view as coronal mass ejections. One comet that hit the Sun was at first thought to lead to a splash of cometary material. It is currently thought that, given the common appearance of coronal mass ejections, a CME coincidentally occurred right after the comet hit. These coincidences were verified dramatically by the independent coronal mass ejection that occurred as a bright comet passed the Sun in the LASCO field of view in 2003.

For decades, scientists have assumed that solar flares—seen shining brightly on the Sun, at least in hydrogen light—caused the magnetic storms on Earth that we discuss in the following section. But in recent years, we have realized that coronal mass ejections are also major contributors to geomagnetic storms.

Indeed, a debate has raged over whether coronal mass ejections are begun by solar flares. The evidence is ambiguous, though I would bet that the resolution of the discussion is that the link is not always there. Sometimes CMEs can erupt without a specific flare occurring.

Usually, we notice coronal mass ejections when the gas sticks out to the side. That is, the eruption is in the plane of the sky, from our point of view. But occasionally, a halo event occurs. In these halo events, we see fairly uniform brightening around the space coronagraph's occulting disk. This symmetry tells us that the coronal mass ejection is pointed at us. Now, the magnetic field in interplanetary space curves around, so even coronal mass ejections that go off to the side can still hit us. But the halo events seem particularly relevant to magnetic storms on Earth.

Coronal mass ejections are important not only for where they are going, but also for where they have been. By carrying magnetic flux away from the Sun, they affect the underlying solar activity. They leave the lower levels of the Sun with a weaker magnetic field.

Space Weather

For thousands of years, people have looked up at the night sky and occasionally seen aurorae. The aurora borealis is the "northern lights," the patterns of glowing atmosphere caused by incoming particles from the Sun. The aurora australis is the similar phenomenon in the southern hemisphere. The aurorae are perhaps the best known type of space weather.

Space weather is the condition of space between the Earth and the Sun, with x-rays and gamma rays flashing by at the speed of light and high-energy particles from the Sun making their way more slowly. Just as people on Earth can skid off the road when

The Solar Scoop

Auroras have been seen on Jupiter and on Saturn, as well as on Earth. When an aurora occurs in one hemisphere, it occurs in the other hemisphere as well. Particles from the Sun are funneled down into the atmosphere symmetrically along magnetic lines of force to circles around both poles.

the weather on Earth is bad, satellites and people in space can be affected or harmed by space weather. The United States' National Oceanographic and Atmospheric Administration (NOAA) has stepped up its forecasting role and its study of the factors that go into space weather.

Some of space weather is the result of the normal solar wind. A correlation has been established between the presence of coronal holes on the Sun and geomagnetic storms, disruptions of the Earth's magnetic field. Since coronal holes represent open field lines at the Sun, the solar wind is relatively free to escape there.

But more violent actions take place with coronal mass ejections and solar flares. Flares have long been known to affect communications on the Earth. Static that occurs soon after flares can wipe out contacts. A lot of research, both military and nonmilitary, has focused on predicting the occurrences of flares. No definitive predictor has ever been found, though astronomers continue to look at some particular configurations of the magnetic field. Astronomers are better at predicting when there will be no flare. If the Sun looks quiet and the sunspot regions are docile and nicely arranged with one polarity on one side and the other polarity on the other side, a solar flare is unlikely. But if you see a jumble of polarities, reconnection and a flare are possible, even though we can't predict just when this would occur.

Coronal mass ejections are so substantial that they batter the Earth's magnetosphere and upper atmosphere. During the solar maximum of 1980, there were few satellites in space to be bothered by coronal mass ejections, but by 1990 there were dozens. And in the most recent solar maximum, around 2001, hundreds of space satellites were vulnerable to coronal mass ejections. Indeed, soon after one halo event occurred, one of the communication satellites went dead. Though nobody has been definitive about it, it sure looks as though there is a link between that dead satellite and the coronal mass ejection.

Even a few minutes' notice of the arrival of solar particles can allow safety measures to be taken. For example, high voltage on satellites can be diminished, making it less likely that the satellites would be disabled. Solar storms reaching the Earth can also cause fluctuations in power lines, sometimes tripping circuit breakers. A few minutes' notice to electric utilities can be helpful here, too.

Astronauts in space shuttles are vulnerable to high-energy radiation. Too much of it can lead to radiation sickness. Predicting when a storm of solar particles will arrive might allow astronauts to retreat into the most ked regions of their spacecraft. The prediction of a major storm might even lead to the abortion of a crewed mission for the protection of the astronauts.

It has gotten easier to monitor space weather from your own home. Many people check www.spaceweather.com on the web. The site is controlled by NOAA, which also sponsors the Sun-Earth Connection (SEC) site at sec.noaa.gov. In addition to general astronomy news, spaceweather.com shows what the Sun looks like recently as seen through many different filters. It shows sunspots on both the near side of the Sun and, newly, the far side. It tells you about coronal holes, solar flares, and geomagnetic storms. And it often gives a prediction of the likelihood of an aurora. So you might add

spaceweather.com to, perhaps, weather.com whenever you go outdoors—or if you are responsible for communications or television relay satellites.

Our Sun is our friend, but it can turn on us. We study the Sun both to love it and to fear it. A panoply of telescopes on the ground and in space—ranging in wavelength coverage from gamma rays to radio waves—is there to protect us and to help us learn about our nearest star and its relation to our home planet.

 Solar Scribblings

Space weather may be a limiting factor in sending people to Mars. Though our Apollo astronauts reached the Moon in four days, the trip to Mars is much longer and will take years. Astronauts could be harmed by particles from flares or coronal mass ejections during that period.

The Least You Need to Know

- ◆ Solar flares are abrupt explosions on the Sun, often reaching tens of millions of degrees.

- ◆ Coronal mass ejections carry huge amounts of matter into interplanetary space.

- ◆ Coronal mass ejections can coincidentally occur when comets are seen near the Sun.

- ◆ Space weather can easily be followed online.

Appendix A

Glossary

annular eclipse   An eclipse of the Sun in which an annulus (a ring) of bright everyday sunlight remains visible around the Moon.

annular-total eclipse   An eclipse of the Sun that is annular for the ends of its path, but total in the middle.

anomalistic month   The period between perigees, 27.55 days—the closest approach of the Moon to the Earth.

antumbra   The continuation of the umbral cone beyond its point.

atom   The smallest building block of the chemical elements. A chemical element is one type of atom.

AURA   Associated Universities for Research in Astronomy, the organization that runs the National Solar Observatory, the National Optical Astronomy Observatories, the Gemini Observatory, and the Space Telescope Science Institute.

black-drop effect   The appearance of a dark band joining Venus's silhouette and the sky during transits, preventing their accurate timing.

butterfly diagram   Graph that shows butterfly shapes when the latitude of sunspots is graphed over time.

carbon cycle   A way to fuel stars by adding hydrogens to a carbon atom and its intermediate transformations. It operates in stars hotter than the Sun.

celestial sphere   The imaginary sphere surrounding Earth, with the stars on it.

central eclipse   A total or annular solar eclipse.

CFCs   *See* chlorofluorocarbons.

chlorofluorocarbons (CFCs)   Compounds made of chlorine, fluorine, and carbon that were thought to be stable but that actually disintegrate in the stratosphere and affect the ozone layer.

chromosphere   The colorful shell around the Sun that pops into view at the beginning and the end of a total eclipse.

CME   *See* coronal mass ejection.

corona   The pearly white crown of light visible around the Sun at a total eclipse.

coronagraph   A type of telescope that makes artificial eclipses. The term has been generalized to include mechanism on other telescopes that block bright parts of an image to allow fainter parts to be seen.

coronal hole   A relatively dark region of the corona seen in x-rays or extreme ultraviolet. Coronal holes represent open regions of the coronal magnetic field, from which gas can easily escape into interplanetary space.

coronal mass ejection (CME)   The eruption and departure from the Sun of a piece of the corona. These ejections, often called CMEs by professionals, have been seen to occur daily near solar maximum daily. They are a major link between the Sun and Earth.

draconic month   The 27.21-day intervals between the Moon's return to the same node. Also called a nodical month.

eclipse season   The time of year when Earth and the Moon are close enough to the nodes to potentially have an eclipse.

eclipse year   The 346.62-day period during which Earth passes through opposite nodes.

electromagnetic radiation   Waves of electricity and magnetism that travel at the speed of light. From shortest to longest wavelengths, it includes gamma rays, x-rays, ultraviolet, visible light, infrared, and radio waves.

ESA   The European Space Agency.

exoplanets   Planets around stars other than our Sun.

extinction   How much of incoming light is absorbed before it reaches us.

extreme ultraviolet   The shortest part of ultraviolet light, farthest from visible light.

filament   What a prominence looks like when you look at it from above, silhouetted against the solar disk.

flare   A powerful, sudden eruption on the Sun, not to be confused with a prominence.

forbidden lines   Spectral lines that would occur so rarely that other circumstances in atoms, like collisions that change the condition of the atoms present, would prevent them from occurring. They appear only in gas of extremely low density.

gamma rays   Electromagnetic radiation, like light but with even shorter wavelengths than x-rays.

global warming   The upward trend in temperature. Most scientists have concluded that most of it comes from the greenhouse effect caused by human contributions of carbon dioxide and other gases to the Earth's atmosphere.

greenhouse effect   When incoming sunlight passes through a transparent atmosphere and is transformed to infrared radiation that is partly blocked from escaping by the atmosphere.

halons   Bromine-based gases that affect the ozone layer.

heliopause   The upper limit of the heliosphere.

helioseismology   The technique of finding out about the inside of the Sun by studying oscillations of the Sun's surface.

heliosphere   The zone in which the Sun has more influence than interstellar space.

H-H objects   Jets of gas given off by young stars.

ion   An atom that has lost one or more of its electrons.

isotopes   Forms of atoms with different numbers of neutrons, which add mass but not charge. For example, the most common isotope of hydrogen has just a proton; another isotope of hydrogen is deuterium, which has a proton plus a neutron.

LWS   NASA's Living with a Star program. Its objectives are (1) to quantify the physics, dynamics, and behavior of the Sun-Earth system over the 11-year solar cycle; (2) to improve understanding of the effects of solar variability and disturbances on terrestrial climate change; (3) to provide data and scientific understanding required for advanced warning of energetic particle events that affect the safety of humans; and (4) to provide detailed characterization of radiation environments useful in the design of more reliable electronic components for air and space transportation systems.

magnetic carpet   Term that shows how common and dynamic changes in the magnetic field are, with a tangled magnetic field found all over the images of the lower corona and a resulting magnetic reconnection on small scales.

magnetic field   Region of influence of a source in which the direction and strength of magnetism can be determined.

magnetic reconnection   The abrupt change of the arrangement of loops in a region on the solar surface, to create a lower-energy and more stable situation.

magnetohydrodynamics   The study of the motion of ionized gases (plasmas) in magnetic fields.

MHD   *See* magnetohydrodynamics.

Mira   The first variable star to be discovered.

moss   A thin layer marking the interface between the corona and the chromosphere; it is 500,000 to 2,000,000°C and changes with the magnetic field.

NASA   The National Aeronautics and Space Agency, the United States space and aeronautics agency.

negative hydrogen ion   A hydrogen atom with a second electron, either loosely bound or merely affected by the hydrogen atom. The radiation from the Sun in the visible and infrared comes primarily from the negative hydrogen atom.

neutrino   A subatomic particle with very tiny mass and no electric charge that travels at almost the speed of light and interacts very poorly with other matter.

NOAO   The National Optical Astronomy Observatory, based in Tucson, Arizona, with its main observing station on nearby Kitt Peak. It is supported by the National Science Foundation.

node   A place where two curves cross or a wave doesn't change over time. Between two orbits, it is the pair of locations where the two orbits cross.

nodical months   The 27.21-day intervals between the Moon's return to the same node. Also called draconic months.

NRL   The U.S. Naval Research Laboratory, based in Washington, D.C.

NSF   The National Science Foundation, which funds most ground-based astronomy in the United States.

NSO   The National Solar Observatory, with stations on Sacramento Peak in New Mexico and Kitt Peak in Arizona. It is supported by the National Science Foundation.

opacity   A measure of how opaque, not transparent, a gas is.

ozone hole   A thinning of the ozone seen in the Antarctic region each southern-hemisphere spring. It results from the breakdown of ozone molecules by chlorofluorocarbons and halons.

penumbra At an eclipse, the set of places that are only partially shadowed. Also, for a sunspot, the filamentary, less-dark region surrounding the umbra.

permitted lines Spectral lines that are common (we say "allowed") from transitions between two energy levels of atoms.

photosphere The surface that we see in visible light.

plasma A gas composed entirely of ions and the electrons that balance the charge.

prominences Structures held in space above the Sun by the Sun's magnetic field and seen when they are on the edge of the Sun.

proton-proton chain The series of nuclear reactions that fuels the Sun.

quiet Sun The everyday, unchanging background Sun.

revolving Orbiting another body.

rotating Spinning on its axis.

saros The interval of 18 years 11⅓ days (plus or minus a day on the calendar, depending on leap years) over which eclipses repeat.

SDO NASA's planned Solar Dynamics Observatory.

SEC NASA's Sun-Earth Connection program.

seeing The quality that describes how steady images are.

seismic waves Vibrations that travel through Earth.

seismology The study of seismic waves, usually on Earth.

seismometers Devices on the Earth or the Moon, and potentially on other planets or moons with solid surfaces, used to measure seismic waves.

SOHO NASA and ESA's Solar and Heliospheric Observatory, launched in 1995.

solar-activity cycle The sunspot cycle matched in other solar phenomena.

Solar-B Japan's next solar spacecraft, which has major U.S. and U.K. participation. Its launch is planned for 2007.

solar constant The amount of energy that reaches a square meter of the top of Earth's atmosphere each second.

solar flare An abrupt brightening of part of the Sun within seconds, including the release of electromagnetic radiation and particles. It is a powerful and sudden eruption, releasing energy stored in the magnetic field.

Solar Orbiter   An ESA spacecraft planned for 2010.

Solar Polar Imager   A wished-for NASA spacecraft, for the distant future.

Solar Probe   A wished-for NASA spacecraft, now in abeyance.

solar wind   The corona expanding into space.

SOLIS   Synoptic Optical Long-term Investigations of the Sun, an American long-term program based at the National Solar Observatory, with a new telescope on Kitt Peak.

space weather   Earth's environment in space as the result of emissions of particles and electromagnetic radiation from the Sun.

spectrograph   A device that records the spectrum on film or otherwise, currently onto digital media.

spectroscope   A device that you can look through to see the spectrum.

STEREO   NASA's planned Solar-TErrestrial RElations Observatory.

sunspot cycle   The approximately 11-year up-and-down in the number of sunspots.

Sun-synchronous orbit   An orbit in which the spacecraft goes around Earth every two hours or so, keeping its orbital plane perpendicular to the direction toward the Sun. This puts the spacecraft in continuous view of the Sun for most of each year.

syzygy   A lineup of three astronomical bodies.

TRACE   NASA's Transition Region And Coronal Explorer, launched in 1999.

transit   The passage of one celestial body in front of another. A total eclipse is a transit of the Moon in front of the Sun.

transition region   The location between the chromosphere and the corona. Solar gas rises in temperature from about 18,000°F to 1,800,000°F (10,000°C to 1,000,000°C) there.

transparency   How clear the atmosphere is.

T Tauri stars   Young stars that are still unsteady in brightness, resembling the early years of the Sun's life.

total eclipse   When the Moon entirely hides the Sun from a part of Earth.

ultraviolet   Electromagnetic radiation just shorter than visible light.

umbra   At an eclipse, the completely dark part of a shadow. For a sunspot, the dark central region with the strongest magnetic field.

x-rays High-energy electromagnetic radiation, like light, but hundreds or thousands of times shorter in wavelength.

Yohkoh Japan's Solar-A spacecraft, which observed x-rays from 1990 for almost 10 years.

zenith The point in the sky directly over your head.

Online Solar Glossaries

www.sunspot.noao.edu/PR/zoo.html

web.ngdc.noaa.gov/stp/GLOSSARY/glossary.html

www.hao.ucar.edu/public/education/glossary.html

solar-center.stanford.edu/gloss.html

pluto.space.swri.edu/IMAGE/glossary/glossary_intro.html

sohowww.nascom.nasa.gov/explore/glossary.html

vestige.lmsal.com/TRACE/Science/ScientificResults/trace_cdrom/html/glossary.html

Solar Observatories

Latest Solar Images

Big Bear Solar Observatory

www.bbso.njit.edu/cgi-bin/LatestImages

Big Bear Solar Observatory Active-Region Monitor

www.bbso.njit.edu/arm/latest/

Solar and Heliospheric Observatory (SOHO)

sohowww.nascom.nasa.gov/data/realtime-images.html

Global High-Resolution H-Alpha Network

www.bbso.njit.edu/Research/Halpha

GOES Solar X-Ray Imager

www.ngdc.noaa.gov/stp/stp.html

www.sec.noaa.gov/sxi

San Fernando Solar Observatory

www.csun.edu/sfo

Rome Observatory

www.mporzio.astro.it/solare

Lockheed Master List of Solar Websites

www.lmsal.com/solarsites.html

Eclipses

Working Group on Solar Eclipses of the International Astronomical Union

www.solarcorona.org

www.williams.edu/astronomy/eclipses

Program Group on Public Education at the Times of Eclipses (of the Commission on Education and Development of the International Astronomical Union)

www.solareclipses.info

Espenak/NASA/Goddard Eclipse Home Page

sunearth.gsfc.nasa.gov/eclipse

Espenak Atlas of Solar Eclipse Paths

sunearth.gsfc.nasa.gov/eclipse/SEatlas/SEatlas.html

Eclipse Maps 2000–2030

www.skylook.net/eclipses/futuro/ms0030.htm

Space Satellites

Solar and Heliospheric Observatory

sohowww.nascom.nasa.gov/

SOHO gallery

sohowww.nascom.nasa.gov/gallery

SOHO: The Sun as Art

sohowww.nascom.nasa.gov/hotshots/2002_08_29/Artsun02lo.pdf

Transition Region and Coronal Explorer

vestige.lmsal.com/TRACE/

Ulysses Probe Homepage

ulysses.jpl.nasa.gov/

Ramaty High Energy Solar Spectroscopic Imager (RHESSI)

hesperia.gsfc.nasa.gov/hessi/outreach.htm

hessi.ssl.berkeley.edu

Imager for Magnetopause-to-Aurora Global Exploration (IMAGE)

image.gsfc.nasa.gov

Thermosphere-Ionosphere-Mesosphere-Energetics and Dynamics (TIMED)

www.timed.jhuapl.edu

GOES Solar X-ray Imager

www.ngdc.noaa.gov/stp/stp.html

www.sec.noaa.gov/sxi

science.msfc.nasa.gov/ssl/pad/solar/sxi.htm

NASA Missions, with links

spacescience.nasa.gov

NASA's Living with a Star Program

lws.gsfc.nasa.gov

Soviet-Ukranian CORONAS spacecraft

coronas.izmiran.rssi.ru/news

Future Solar Satellites and Programs

Solar-B

wwwssl.msfc.nasa.gov/ssl/pad/solar

Solar Dynamics Observatory

sdo.gsfc.nasa.gov

Solar Probe

www.jpl.nasa.gov/ice_fire/sprobe.htm

Solar Orbiter

spdext.estec.esa.nl/content/doc/42/24386_.htm

Solar TErrestrial RElations Observatory (STEREO)

stereo.jhuapl.edu

Synoptic Optical Long-term Investigations of the Sun (SOLIS)

solis.nso.edu

Past Spacecraft

Yohkoh Public Outreach Project (includes photos and movies)

www.lmsal.com/YPOP/

Solar Maximum Mission

www.hao.ucar.edu/public/research/svosa/smm/smmcp_cme.html

Solar Observatories

Big Bear Solar Observatory

www.bbso.njit.edu/

Swedish Solar Observatory, La Palma

www.solarphysics.kva.se

National Solar Observatory

www.nso.edu

www.noao.edu/image_gallery/solar.html

Global Oscillation Network Group (GONG)

gong.nso.edu

Stanford SOLAR Center

solar-center.stanford.edu

High Altitude Observatory

www.hao.ucar.edu/public/education/education.html

Mauna Loa Solar Observatory

mlso.hao.ucar.edu/

Advanced Technology Solar Telescope

atst.nso.edu

Solar Web Information

Author's Eclipse Expeditions

www.williams.edu/astronomy/eclipse

SEDS (Students for the Exploration and Development of Space) Home Page

www.seds.org/nineplanets/nineplanets/sol.html

From Stargazers to Starships

www.phy6.org/stargaze

Exploratorium

www.exploratorium.edu/observatory

Multimedia Tour

www.astro.uva.nl/demo/od95

Eric Weisstein's World of Astronomy

scienceworld.wolfram.com/astronomy/Sun.html

Sun-Earth Connection Forum

sunearth.gsfc.nasa.gov

Sunspot Numbers

Sunspot Numbers Online

sidc.oma.be

Sunspots and Butterfly Diagram

wwwssl.msfc.nasa.gov/ssl/pad/solar/sunspots.htm

Solar Constant

World Radiation Center/Physical Meteorological Observatory Davos

www.pmodwrc.ch, under Projects

Space Weather

spaceweather.com

spaceweather.com

NOAA Space Environment Center

sec.noaa.gov

Lockheed Master List of Solar Websites

www.lmsal.com/solarsites.html

NOAA Data and Links

www.sel.noaa.gov/solcoord/solcoord.html

Magnetosphere

www.spof.gsfc.nasa.gov/Education/Intro.html

Auroras

www.gi.alaska.edu

University Corporation for Atmospheric Research

www.windows.ucar.edu/spaceweather/basic_facts.html

Appendix C

Astronomy Clubs and Solar Interest Groups

Amateur Societies with Solar Divisions

Association of Lunar & Planetary Observers (ALPO)

www.lpl.arizona.edu/~rhill/alpo/solstuff/solinks.html

The ALPO Solar website is at www.lpl.arizona.edu/~rhill/alpo/solar.html

American Association of Variable Star Observers

25 Birch Street
Cambridge, MA 02138
www.aavso.org/solar

The association's Solar Committee compiles the American Relative Sunspot Numbers.

General Astronomy Clubs

For your local Club, see www.skyandtelescope.com. Choose Resources, Clubs and Organizations from the main page.

Amateur Astronomers Association of New York

www.aaa.org

Astronomical League

The Astronomical League is the umbrella group of amateur societies. For its newsletter, *The Reflector*, write:

The Astronomical League
Executive Secretary
c/o Science Service Building
1719 N St., N.W.
Washington, DC 20030
www.astroleague.org

Astronomical Society of the Pacific

390 Ashton Ave.
San Francisco, CA 94112
www.astrosociety.org

British Astronomical Association

Burlington House
Piccadilly, London W1V 0NL, England
www.britastro.org

The Solar Section is at baa-solarsection.org.uk.

Royal Astronomical Society of Canada

124 Merton St.
Toronto, Ontario M4S 2Z2, Canada
www.rasc.ca

Amateur Solar Images

Art Whipple

www.chesapeake.net/~osprey/sunspots.html

Jack Newton

www.jacknewton.com

Eclipse Images

eclipse.span.ch/total.htm

solareclipsewebpages.users.btopenworld.com/

www.mreclipse.com

www.williams.edu/astronomy/eclipse

A Word on Temperature

In the United States, we mainly use Fahrenheit degrees, a scale on which water freezes at 32° and boils at 212°, some 180° hotter. In 1742, the Swedish astronomer Andreas Celsius suggested a temperature scale that seemed more rational: Water freezes at 0° and boils at 100°. That scale used to be called centigrade but now is named Celsius.

Kelvin Temperature Scale

The Sun has no freezing or boiling water, though, so neither the Fahrenheit nor the Celsius scale is particularly sensible. We want to use a scale that starts at as low a temperature as possible, which we will call "zero." Just think of "absolute zero" as the lowest temperature that could be conceived of. Astronomers use the kelvin temperature scale, which uses degrees that are the same size as Celsius degrees but that begins at absolute zero. In the current system of units that has been adopted internationally, the intervals are actually called "kelvins" rather than "degrees kelvin" or "°K." In this adopted system, units named after people have capital letters in their symbols. So, the symbol for a kelvin is K. It is named after Lord Kelvin, the English scientist who started out as William Thomson but was named Baron Kelvin. He worked out the kelvin scale in 1848.

Absolute zero has been measured to be –273.16°C. So on the kelvin scale, water freezes at about 273 kelvins. Water boils exactly 100 kelvins higher, at about 373 kelvins. But all these temperatures are much too low to be useful in talking about stars.

The Sun's surface is about 5,800 kelvins. Subtract 273 to find its Celsius temperature. Since we are talking in round numbers, and 273 is about 300, let's say that the Sun's surface is about 5,500°C.

To change from Celsius temperatures to Fahrenheit temperatures, first multiply by $\frac{9}{5}$, since Celsius degrees are $\frac{9}{5}$ times greater than Fahrenheit degrees. (180 °F = 100 °C, and $\frac{180}{100} = \frac{18}{10} = \frac{9}{5}$). Then just add 32 to get the Fahrenheit temperature. (I always test with 0°C to see that I am doing things right: $0° \times \frac{9}{5} = 0°$. Then add 32° to get 32°F, which is right for the freezing point of water.)

So to change from the Sun's 5,500°C to Fahrenheit, we multiply $5,500 \times \frac{9}{5} = 1,100 \times 9 = 9,900$ and then add 32. That gives us about 10,000°F. Don't worry about a lot of digits or extra decimals. When we are talking about 10,000°, what's a few degrees more or fewer among friends?

Celsius to Fahrenheit

What's an easy way to remember how to change from Celsius to Fahrenheit? Remember the rhyming jingle: "Times 2, minus point 2, plus 32." That is, multiply the Celsius temperature by 2. To find "point 2" (0.2) of the original value, just move the decimal place to the left by one notch. (0.2 is 2.0 with the decimal point moved left.) Then add 32. See the following examples:

- **Terrestrial example:** A nice day at home might be 30°C. Times 2 is 60; then subtract 6 to give 54. Finally, add 32 to get 86.

- **Solar example:** The Sun is 5,500°C at its surface. That times 2 is 11,000. Subtract 1,100 to get 9,900. Then add 32 to get 9,932, or almost 10,000°F. That's close enough.

Selected Readings

See the author's books at www.solarcorona.com.

Books About the Sun in General

Brody, Judit. *The Enigma of Sunspots: A story of discovery and scientific revolution.* Edinburgh: Floris Books, 2002.

Carlowicz, Michael J., and Ramon E. Lopez. *Storms from the Sun: The Emerging Science of Space Weather.* Washington, D.C.: Joseph Henry Press, 2002.

Golub, Leon, and Jay M. Pasachoff. *Nearest Star: The Exciting Science of Our Sun.* Cambridge: Harvard University Press, 2001, 2002. A non-technical, illustrated, trade book. www.williams.edu/astronomy/neareststar.

———. *The Solar Corona.* New York and Cambridge, U.K.: Cambridge University Press, 1997. An advanced textbook. www.williams.edu/astronomy/corona.

Lang, Kenneth R. *The Cambridge Encyclopedia of the Sun.* Cambridge: Cambridge University Press, 2001.

Odenwald, Sten F. *The 23rd Cycle: Learning to Live with a Stormy Star.* New York: Columbia University Press, 2001.

Stix, Michael. *The Sun.* New York, Heidelberg, and Berlin: Springer, 1989.

Zirker, Jack. *Journey from the Center of the Sun.* Princeton: Princeton University Press, 2001.

———. *Sunquakes: Probing the Interior of the Sun.* Baltimore: The Johns Hopkins University Press, 2003.

Author's Textbooks

Pasachoff, Jay M. *Astronomy: From the Earth to the Universe.* 6th ed. Belmont, Calif.: Brooks/Cole Publishing, 2002. info.brookscole.com/pasachoff, www. solarcorona.net.

Pasachoff, Jay M., and Alex Filippenko. *The Cosmos: Astronomy in the New Millennium.* 2nd ed. Belmont, Calif.: Brooks/Cole Publishing, 2004. info.brookscole.com/ pasachoff.

Books on Eclipses

Brunier, Serge, and Jean-Pierre Luminet. *Glorious Eclipses: Their Past, Present and Future.* Cambridge, U.K., and New York: Cambridge University Press, 2000.

Espenak, Fred. *Fifty Year Canon of Solar Eclipses, 1986–2035.* Greenbelt, Md.: NASA Reference Publication 1178, 1987.

Guillermier, Pierrer, and Serge Koutchmy. *Total Eclipses: Science, Observations, Myths and Legends.* Chichester, U.K., and New York: Springer-Praxis, 1999.

Harrington, Philip S. *Eclipse! The What, Where, When, Why & How Guide to Watching Solar and Lunar Eclipses.* New York: Wiley, 1997.

Littmann, Mark, Ken Willcox, and Fred Espenak. *Totality: Eclipses of the Sun.* 2nd ed. New York and Oxford, U.K., and New York: Oxford University Press, 1999.

Maunder, Michael, and Patrick Moore. *The Sun in Eclipse.* New York, Heidelberg, and Berlin: Springer, 1998.

Ottewell, Guy. *The Under-Standing of Eclipses*. Greenville, S.C.: Astronomical Workshop, 1991.

Pasachoff, Jay M., and Michael A. Covington. *The Cambridge Eclipse Photography Guide*. New York and Cambridge, U.K.: Cambridge University Press, 1993.

Steel, Duncan. *Eclipse*. Washington, D.C.: John Henry Press, 2001.

Zirker, Jack B. *Total Eclipses of the Sun*. Princeton: Princeton University Press, 1995.

Traditional Solar Readings

Menzel, Donald H. *Our Sun*. 2nd ed. Cambridge: Harvard University Press, 1959.

Noyes, Robert W. *The Sun: Our Star*. Cambridge: Harvard University Press, 1982.

Observing Reference Books

Espenak, Fred, and Jay Anderson. *NASA Technical Publications*. A series of *NASA Technical Publications*, one for each eclipse or group of eclipses. See http://sunearth.gsfc.nasa.gov/eclipse/SEpubs/bulletin.html

Observer's Handbook (yearly). Royal Astronomical Society of Canada, 136 Dupont Street, Toronto, Ontario M5R 1V2 Canada.

Pasachoff, Jay M. *A Field Guide to the Stars and Planets*. 4th ed. Boston: Houghton Mifflin Co., 2000. All kinds of observing information, including monthly maps and the 2000.0 sky atlas by Wil Tirion, and Graphic Timetables to locate planets and special objects like clusters and galaxies. See www.williams.edu/astronomy/fieldguide.

————. *Peterson's First Guide to Astronomy*. Boston: Houghton Mifflin Co., 1997. A brief, beautifully illustrated introduction to observing the sky. Tirion monthly maps.

————. *Peterson's First Guide to the Solar System*. Boston: Houghton Mifflin Co., 1997. Color illustrations and simple descriptions mark this elementary introduction. Tirion maps of Mars, Jupiter, and Saturn's positions through 2010.

Monthly Non-Technical Magazines in Astronomy

Sky & Telescope, 49 Bay State Road, Cambridge, MA 02138, 1-800-253-0245, www.skyandtelescope.com.

Astronomy, 21027 Crossroads Circle, P.O. Box 1612, Waukesha, WI 53187, 1-800-533-6644, astronomy.com.

Mercury, Astronomical Society of the Pacific, 390 Ashton Ave., San Francisco, CA 94112, 415-337-1100, www.astrosociety.org.

StarDate, 2609 University, Rm. 3.118, University of Texas, Austin, TX 78712, 1-800-STARDATE, stardate.utexas.edu.

The Griffith Observer, 2800 East Observatory Road, Los Angeles, CA 90027, www.griffithobs.org.

Careers in Astronomy

American Astronomical Society Education Office, 2000 Florida Ave., NW, Suite 400, Washington, D.C. 20009, fax: 202-234-2560; aased@aas.org; www.aas.org/education. A free booklet, *Careers in Astronomy*, is available online.

Movie and Video

SolarMax, made as an IMAX movie, is available on DVD and VHS videotape. This 40-minute movie shows solar images from the ground and from space and is available at many science-museum bookstores and over the Internet: www.solarmovie.com.

Index

A

A-bomb (atomic bomb), 35
AAVSO (American Association of Variable Star Observers), 89
Abbot, Charles Greeley, 282-283
absorption lines, exoplanets, 58-60
ACRIM (Active Cavity Radiometer Irradiance Monitor), 232, 284
Active Cavity Radiometer Irradiance Monitor. *See ACRIM*
Adams, John, 98
Advanced Technology Solar Telescope. *See ATST*
American Association of Variable Star Observers. *See AAVSO*
American Astronomical Society, 240
American eclipses
  annular, 152-153
  future, 160-161
  partial eclipses, 154
American expeditions, viewing transits of Venus, 175
analemmas, 45-47
Anasazi tribe, monuments as observatories, Chaco Canyon sun dagger, 83
angstroms, 197, 262-263
angular momentum, 64-65
anisotropies, 255
annular eclipses, 125
  American eclipses, 152-153
  future, 159
annular-total eclipses, 125
anomalistic months, 129
antumbra, 127

Apollo, Apollo Telescope Mount, 229-231
archaeoastronomers, 82-83
argon atoms (detecting neutrinos), Davis, Ray, 36-37
Aristotle, stationary Earth theory, 91
astronomers, U.S. National Academy of Sciences, 274
Astrophysical Observatory of the Smithsonian Institution, 282
atmosphere, 4
  chromosphere, 7
  heliosphere, heliopause, 10
  photosphere, 6
  solar energy, 294-304
  study of Sun from outer-space, 228-231
  temperature of the Sun's layers, 8
  Venus, 176
atomic bomb (A-bomb), 35
atoms, 34, 66
  electrons, 66
  elements in the Sun, 68
  helium, 67
  hydrogen, 66-67
  ions, plasma gases, 66
  isotopes, 67
  neutral state, 66
  neutrons, 66
  nuclei, 34, 66
  protons, 66
ATST (Advanced Technology Solar Telescope), 210-213
Aubrey holes, as predictor of solar and lunar eclipses, 75
Aubrey, John, 75
aurorae, space weather, 311-313

autumnal equinox, 42-43
Aveni, Anthony, 83
azimuth, 42
Aztec civilization, monuments as observatories, 82

B

Bahcall, John, 36-37
Balmer series, spectral lines, 69
Bayer, Johann, 89
Beijing Astronomical Observatory, 207
Bethe, Hans, carbon cycle, 33-34
Bhatnagar, Arvind, 208
Big Bear Solar Observatory, 190, 201
Bighorn Medicine Wheel, 83
black polarity, 21
black-body curves, 85-87
black-drop effect
  transits of Mercury, 179
  transits of Venus, 173
blue sky, Rayleigh scattering, 53
blurring (seeing property of images), 188
Bohr, Neil, 144
bolometers, measuring solar constant, 282-283
bow shock (planets), 10
Brahe, Tycho, 95
brown dwarfs, 71
Bruno, Giordano, 59, 100
Buhl Planetarium, siderostat telescopes, 196
Butler, Paul, 100
butterfly diagrams (Maunder), 23

C

C-class solar flares, 309
C. E. K. Mees Solar
  Observatory, 198-199
calendars, 49-50
cameras, lenses, 163-164
Canaries Institute for
  Astronomy and Astrophysics,
  206
Canary Islands, 205
Cannon, Annie Jump, 71
carbon cycle, 34
carbon dioxide, greenhouse
  effect, 295-298
Carhenge, 75
cathedrals as observatories, 79-80
CCDs (charge-coupled devices),
  165, 222
CDS (Coronal Diagnostic
  Spectrometer), 254
celestial equator, 42
celestial sphere, 40
CELIAS (Charge, Element, and
  Isotope Analysis System), 257
central eclipses, 125
centrifugal force, 64
Cepheid variable stars, 89
CFCs (chlorofluorocarbons),
  301-304
Chaco Canyon sun dagger, 83
Challenger, 233
Chamorro time zone, 47
Chandra X-ray Observatory,
  110, 239
Chandrasekhar, Subrahmanyan,
  109
Charge, Element, and Isotope
  Analysis System. *See* CELIAS
charge-coupled devices. *See*
  CCDs
Charon moon (Pluto), 99
China, Chinese Academy of
  Sciences, Beijing Astronomi-
  cal Observatory, 207
chlorine atoms, detecting neutri-
  nos, Davis, Ray, 36-37

chlorofluorocarbons. *See* CFCs
chromosphere, 7
  total eclipses, 118
civilization, 73-82
climate, solar constants, 290-291
CMEs (coronal mass ejections),
  251-252, 310-311
coal (fossil fuel), greenhouse
  effect, 298
codices, 82
color of the Sun, 51-52
  green flash, 54-55
  radio waves, 57
  rainbows, 55-60
  Rayleigh scattering, 53
  red sunsets, 53
  Sun dogs, 56
  Sun pillars, 56-57
color-magnitude diagrams, 103
comets, Hale-Bopp, 10
Comprehensive SupraThermal
  and Energetic Particle. *See*
  COSTEP
cone of the Moon's shadow,
  eclipses, 130-131
cones, 52
constellations, Corona Borealis,
  89
converging series, spectral lines,
  69
Cook, James, 172
Copernicus, Nicolaus
  Copernican theory, 80
  Sun at the center of the uni-
    verse, retrograde motion,
    92-93
cornal holes, 253-254
corona, 7
  changing shape with the
    sunspot cycle, 120-121
  CMEs (coronal mass ejec-
    tions), 310-311
  coronagraphs, 8-9
  emission lines, 145
  presence of coronium,
    140-141
  solar wind, 9
  space streamers, 142

  visibility during a total solar
    eclipse, 115-119
  wind, 255
Corona Borealis, 89
coronagraphs, 8-9, 199
  LASCO (Large Angle
    Spectroscopic
    Coronagraph), 9
  Mauna Loa Observatory, 200
  SOHO, 249-252
Coronal Diagnostic
  Spectrometer. *See* CDS
coronal loops, moss, 267-268
coronal mass ejections. *See*
  CMEs
coronium, presence in corona,
  140-141
cosmic rays, 36
COSTEP (Comprehensive
  SupraThermal and Energetic
  Particle), 257
Crab Nebula, 108
Crutzen, Paul, 302
Curie, Marie, 67

D

Daguerre, Louis J. M., 138
Danielson, Dennis, 92
data, solar, GONG+, 223
Davis, Ray, detection of neutrinos, 36-37
daylight saving time, 48
De la Rue, Warren, 138
declination, 42
degeneracy, electron, 109
detecting neutrinos
  Davis, Ray, 36-37
  GALLEX, 37
  Kamiokande (Kamioka
    Neutrino Detection
    Experiment), 37
  SAGE (Soviet-American
    Neutrino Experiment), 37
detecting sunspots, 77-79
deuterium, 67

deuterons, 67-68
di Cicco, Dennis, 46
diamond-ring effect, total eclipse, 118
DIARAD (Differential Absolute RADiometer), 286
Differential Absolute RADiometer. See DIARAD
differential rotation of the Sun, leading sunspots versus trailing sunspots, 19-21
digital cameras, photographing eclipses (megapixels), 165
Division of Planetary Sciences of the American Astronomical Society, 178
Dobson units, measuring ozone in atmosphere, 304
Dobson, G. M. B., 304
Doppler, Christian, 99
Doschek, George, 271
double rainbows, 56
draconic months, 127
Dresden Codex, 82
Dumbbell Nebula, 106
Dunn, Richard B., 197
duration of sunlight, 44
dust, formation of the Sun, 63-67
dwarf stars, 105-106
dynamics (study of things moving), 143

E

e (irrational number), 28
Eagle Nebula, 65
Earth
    earthquakes, seismology, 30
        seismic waves, 217
    elliptical orbit, 44
    syzygy, 127
eclipse years, 126
eclipses, 127
    annular eclipses
        American eclipses, 152-153
        future, 159
        anomalistic months, 129

cone of the Moon's shadow, 130-131
    filters, 131
    historical, 133-138
    partial eclipses, 116
        American eclipses, 154
        filters, 116
    phases of the Moon, 123-125
    rocket data, 240
    saros, 128-130
    seasons, 126-127
    syzygy, 127
    total eclipses, 6, 113-119
    widths and lengths, 132
Eddington, Arthur, 33, 100
Eddy, Jack, 83
Edlén, Bengt (spectroscopist), 140
EGGs (Evaporating Gaseous Globules), 65
Einstein, Albert, 34
EIT (Extreme-ultraviolet Imaging Telescope), 252-254
electromagnetism, 15
electron degeneracy, 109
electronic detectors. See CCDs (charge-coupled devices)
electrons
    atoms, 66
    energy levels, 144-145
elements, atoms, 66-68
    electrons, 66
    examining the Sun's spectrum, 68
    helium, 67
    hydrogen, 66-67
    ions, 66
    isotopes, 67
    neutral state, 66
    neutrons, 66
    nuclei, 66
    proton-proton chain, 68
    protons, 66
elliptical orbit
    Earth, 44
    planets, 96
emission lines, 58

Energetic and Relativistic Nuclei and Electron experiment. See ERNE
energy
    atmospheric warming, 294-304
    equilibrium, 294
    levels, electrons, 144-145
    solar constant, 282
energy-storage mechanism, space weather and solar flares, 307
epicycles, 92
equation of time, 47
equilibrium, solar energy, 294
equinoxes, 41-43
    autumnal, 42-43
    vernal, 41-43
ERNE (Energetic and Relativistic Nuclei and Electron experiment), 257
European eclipses, 151-152
European Space Agency
    Eddington mission, 100
    partnership with NASA, 247
    Solar Orbiter, 276-277
EUV (extreme ultraviolet), 22
Evans, Jack, 199
Evaporating Gaseous Globules. See EGGs
exoplanets, 11
    Fraunhofer lines (absorption lines), 59-60
expeditions, viewing transits of Venus, 174-175
Explorers, TRACE (Transition Region and Explorer Spacecraft), 260-262
    gas loops, 263-265
    magnetic carpet, 266-267
    magnetic reconnection, 265-267
    moss, 267-268
    telescope, 262
    ultraviolet wavelengths, 262
exposure times, camera lenses, 164

extinction, transparency of images, 188
extrasolar planets. *See* exoplanets
extreme ultraviolet. *See* EUV or XUV
Extreme-Ultraviolet Imaging Telescope (SOHO), 30, 252-254

F

Fabricius, David
  detecting sunspots, 79
  discovery of Mira, 88
Fabricius, Johannes, 79
faculae, distribution of solar radiation across the Sun, 287-290
far-side imaging, 220-221
Farman, Joseph, 302
fast wind, 274
fields of view, camera lenses, 163
figure-8s. *See* analemmas
filaments, 23
filters
  camera lenses, 164
  eclipses, 131
  reflection/absorption of light, neutral density, 52
  viewing partial eclipses, 116
fission, 33-35
five-minute oscillation (of solar surface), images, 215-216
flares, Small Explorers, 243
flavors (neutrinos), 37-38
focal length, camera lenses, 163
forbidden lines, coronal emission lines, 144-145
formation of stars, 85
  black-body curves, 85-87
  H-H objects, 87-88
  variable stars, 88-90
formation of the Sun, gas and dust, 63-67
fossil fuels, greenhouse effect, 298
Foucault, Jean Bernard Léon, siderostat telescopes, 196

Fraunhofer, Joseph, spectroscopy, 138
  Fraunhofer lines (absorption lines), 57-60
Freon, 43
  ozone hole, 302
fusion, 33
  carbon cycle, 34
  fueling the Sun, 33-35
  nuclear fusion, tokamaks, 142
future
  annular eclipses, 159
  of the Sun
    electron degeneracy, 109
    giant stars and dwarf stars, 103-106
    nebulae, 106-108
    supernovae, 108
  total eclipses, 157-162
  transits of Venus, 179-180

G

G2 stars, 71
Galileo
  conviction of disobedience, 94
  detecting sunspots, 78
  precursor to law of inertia, 93
  telescopic discoveries, 94
Galle, Johann, 98
GALLEX, detecting neutrinos, 37
Gallium, detecting neutrinos, 37
gamma rays, solar flares, 22
gas (fossil fuel), greenhouse effect, 298
gas loops, 263-268
gaseous state of matter, 66
gases
  formation of the Sun, 63-67
  ionized, plasma, 143
  opacity, 28
  optical thickness, 27
  plasma, 66
Gassendi, Pierre, 177
gauss (G), 17

Geostationary Operational Environmental Satellite. *See* GOES
Giacconi, Riccardo, 231
giant stars, 105
Gingerich, Owen, 80
Global Oscillation Network Group. *See* GONG
Global Oscillations at Low Frequencies. *See* GOLF
gnomon, 47
GOES (Geostationary Operational Environmental Satellite), 245
GOLF (Global Oscillations at Low Frequencies), 255
Golub, Dr. Leon, 140, 178, 260, 271
GONG (Global Oscillation Network Group), 31, 222
GONG+, 222-223
GONG++, 32
Goodricke, John, 89
gravity, formation of the Sun, 63-67
grazing incidence, x-rays, 239
green flash, 54-55
Green, Charles, 172
greenhouse effect
  atmospheric warming, 294-298
  equilibrium, 294
  Industrial Revolution, 298-299
  limiting human contribution, 299-301
  ozone hole, 301-304
Gregorian calendar, 50

H

H-bomb (hydrogen bomb), 35
H-H objects (Herbig-Haro objects), formation of stars, 87-88
H-lines, spectral lines, 70
H-R diagrams (Hertzsprung-Russell diagrams), 103-104

Hale, George Ellery, 20, 189
  Hale-Bopp comet, 10
  Hale's Polarity Law, 21
Haleakala, 198
Halley, Edmond, 170-172
halo orbit, 32
halons, ozone hole, 302-303
Haro, Guillermo, 88
Harriot, Thomas, 79
Hathaway, David, 24
heavy hydrogen, 67
Hecht, Anthony, 93
heliopause, 10
helioseismology, 30-33, 217
  SOHO
    GOLF, 255
    interferometry, 255-256
    solar wind, 257
    VIRGO, 256
  views of far side waves,
    220-221
  wave oscillation, 218-220
heliosphere, 10, 24
helium, 67
  discovery studying the Sun,
    139-140
helmet streamers, 142
Herbig, George, 88
Herbig-Haro objects. *See* H-H
  objects
Herschel, William, 98
Hertzsprung, Ejnar, 103
  Hertzsprung-Russell dia-
  grams. *See* H-R diagrams
Hey, James, 228
High Resolution Solar
  Observatory, 260
historical eclipses, 133-138
Hollander, John, 93
Hubble Space Telescope, Eagle
  Nebula, 65
hydrogen atoms, 66-67
hydrogen bomb (H-bomb), 35

I

ice crystals, Sun dogs and Sun
  pillars, 56
identical telescopes, GONG
  (Global Oscillation Network
  Group), 31
images
  far-side imaging, 220-221
  five-minute oscillation of
    solar surface, 215-216
  N-rays, 238
  size, eclipse videography, 166
  x-rays, 237-245
    GOES, 245
    grazing incidence, 239
    RHESSI, 243-244
    Small Explorers, 243
    telescopes, 239
    Yohkoh, 241-243
India, Udaipur Solar
  Observatory, 208-209
Industrial Revolution, greenhouse effect, 298-299
infrared
  radiation, 22
    atmospheric warming,
      295-298
  solar observations, 191-192
    high sites, 192
    instrumented airplanes,
      192
    lag behind optical obser-
      vations, 191-192
Institute for Solar Physics of
  the Royal Swedish Academy
  of Sciences, 206
Institute of Space and Astronautical Science. *See* ISAS
Interferometry, SOHO, 255-256
Io (Jupiter's moon), volcanism,
  218
iodine cells, 216-217
ionized gases, plasma, MHD
  (magnetohydrodynamics), 143
ions, 66, 254
irrational numbers, 28

ISAS (Institute of Space and
  Astronautical Science), 241
Islas Canarias. *See* Canary Islands
isotopes, 67
isotropic, 255

J

Janssen, Pierre Jules César, 139
Japan
  ISAS (Institute of Space and
    Astronautical Science), 241
  Japanese Alps, Nobeyama
    Radio Observatory, 209-210
  Solar-B, 269-270
    stereo views, 271-273
    telescopes, 271
    ultraviolet wavelengths,
      271
    views, 270-271
    x-ray views, 271
  Yohkoh, 241
    "death of," 242
    science, 242-243
Julian calendar, 50

K

K-lines, spectral lines, 70
Kamioka Neutrino Detection
  Experiment. *See* Kamiokande
Kamiokande (Kamioka Neutrino
  Detection Experiment), 37
Kepler mission (NASA), 100
Kepler, Johannes, 95
  detecting sunspots, 79
  laws of planetary motion, 45,
    96, 170
  orbit of Mars, 96
Kitt Peak, McMath-Pierce
  Telescope, 196
Knowth, Newgrange, 77
Koshiba, Mashatoshi, 37
Koutchmy, Serge, total eclipse,
  149
Krupp, Edward, 83

Kuiper Airborne Observatory, 192
Kyoto Protocol, cutting emissions of greenhouse gases, 300

L

L stars, 71
lake observatories, reducing turbulence, 190-191
Langley, Samuel Pierpont, 282-283
Large Angle Spectroscopic Coronagraph. *See* LASCO
LASCO (Large Angle Spectroscopic Coronagraph), 9
    studying CMEs (coronal mass ejections), 310
    telescopes, SOHO, 249-252
laws of motion, Newton, Isaac, 99
laws of planetary motion, 45
    Kepler, Johannes, 96, 170
layers of the Sun, 7-10
Le Gentil, Guillaume Joseph Hyacinthe Jean Baptiste, 174-175
Le Verrier, Urbain, 98
leading sunspots, 20-21
leap years, 49
Leighton, Robert, 216
lenses (cameras), 163-164
light
    colors of the Sun, 51-60
    radiating invisible light, 29
    visible, 22
    white, 28
limb darkening, transit of Mercury, 179
lines of ionized calcium, spectral lines, 70
liquid state of matter, 66
Lockheed Solar Observatory, 189
Lockyer, Norman, 140
loops (gas), 263-265
    magnetic carpet, 266-267

magnetic reconnect ion, 265-267
moss, 267-268
lunar phases, 123-131
    annular eclipses, 125
    annular-total eclipses, 125
    anomalistic months, 129
    central eclipses, 125
    cone of the Moon's shadow, 130-131
    eclipse seasons, 126-127
    saros, 128-130
    syzygy, 127
Lyot, Bernhard, 199

M

M stars, 71
M-class solar flares, 309
Madrid Codex, 82
magnetic carpet, 266-267
magnetic fields
    MHD (magnetohydrodynamics), 142
    sunspots, 15-17
magnetic reconnection, 265-267
magnetism, 15
    Zeeman effect, 21
magnetohydrodynamics. *See* MHD
Marcy, Geoff, 100
Marshall Space Flight Center, Hathaway, David, 24
mass (neutrinos), 37
Mauna Loa Observatory, 200
Maunder, Walter E., 23-24
Mayan civilization, monuments as observatories, 82
Mayor, Michel, 100
McMath-Pierce Telescope, National Solar Observatory (NSO), 196
mean solar time, 46

measuring
    magnetism, Zeeman effect, 21
    solar constant, 282-287
    solar system, Halley, Edmond, 170-172
medicine wheels, monuments as observatories, 83
Mees Solar Observatory, 198-199
megapixels, photographing eclipses with digital cameras, 165
Menzel, Donald H., 68, 197
Mercury, transits of Mercury, 177
    black-drop effect, 179
    limb darkening, 179
    pairs, 177
metal-deposited filters, 52
methane, greenhouse effect, 298
MHD (magnetohydrodynamics), 142-143
Michelson, Albert, 255
midnight Sun, 43
Milky Way, 63
Mira, 89
Mitchell, Maria, 175
Molina, Mario, 302
monuments as observatories, 82-83
Moon
    phases, 123-131
moss, 267-268
motion, measurements of, iodine cells, 216-217
mountaintop telescopes, 190
Mt. Wilson Observatory, 189
Mylar filters, 52

N

N-rays, 238
NASA
    Apollo program, 229-231
    Chandra X-Ray Observatory, 239
    Eclipse mission, 200

future priorities
  expenses, 277
  SDO, 273-274
  Solar Probe, 274
  Solar-B, 269-270
stereo views, 271-273
telescopes, 271
ultraviolet wavelengths, 271
views, 270-271
x-ray views, 271
  Sun–Earth Connection
  Program, 275
Solar Polar Imager, 275
Kepler mission, 100
monitoring ozone hole, 302
Orbiting Solar Observatory 1
  (OSO-1), 229
partnership with European
  Space Agency, 247
satellites
  SIM (Space Interfero-
  metry Mission), 100
  Terrestrial Planet Finder,
  100
SMM (Solar Maximum
  Mission), 232-233
SOHO, 248-257
  Coronal Diagnostic
  Spectrometer (CDS), 254
  Extreme-ultraviolet
  Imaging Telescope,
  252-254
  GOLF (Global
  Oscillations at Low
  Frequencies), 255
  interferometry, 255-256
  ions, 254
  LASCO telescopes,
  249-252
  science, 249
  solar wind, 257
  UltraViolet Coronagraph
  Spectrometer, 254
  VIRGO, 256
Spacelab, 233

TRACE, 260-268
  gas loops, 263-265
  magnetic carpet, 266-267
  magnetic reconnection,
  265-267
  moss, 267-268
  telescope, 262
  ultraviolet wavelengths, 262
  Ulysses, 234
National Oceanographic and
  Atmospheric Administration.
  See NOAA
National Solar Observatory. See
  NSO
ND (neutral density), 52
nebulae, 106-108
Neptune, discovery, 98
neutral density. See ND
neutral state of atoms, 66
neutrinos, 35-38, 145
neutrons (atoms), 66
Newcomb, Simon, 175
Newton, Isaac, 99
nighttime observatories versus
  solar observatories, 185-189
Nimbus series, measuring solar
  constant, 286
NOAA (National Oceano-
  graphic and Atmospheric
  Administration), 245, 312
Nobeyama Radio Observatory,
  309
nodes, 126
nodical months, 127
nonstellar objects, 106
north star (Polaris), 41
Noyes, Robert, 216
NSO (National Solar
  Observatory), McMath-Pierce
  Telescope, 196
nuclear energy, reducing greenhouse effect, 301
nuclear fusion, 33, 142
  carbon cycle, 34
  tokamaks, 142
nuclear power, 142
nuclei of atoms, 34, 66-67

O

O stars, 71
oblated Sun, 289
oil (fossil fuel), greenhouse
  effect, 298
opacity, gases, 28
optical radiation. See light
optical thickness, 27
orbit of Mars, Kepler data, 96
Orbiting Solar Observatory 1
  (OSO-1), 229
orbits, Sun-synchronous orbit,
  261
oscillations
  GOLF (Global Oscillations
  at Low Frequencies), 255
  interferometry, 255-256
  solar wind, 257
  Sun's surface, helioseismology, 30-32
  VIRGO, 256
  waves, 218-220
Owl Nebula, 106
ozone hole, 43, 301-304

P

p-modes, 218
pairs
  transits of Mercury, 177
  transits of Venus, 172
partial eclipses, 116
  American eclipses, 154
  filters, 116
Payne-Gaposchkin, Cecilia,
  study of spectra, 68
peaks, sunspot cycle, 19
penumbra (sunspots), 14, 127
perigees, 129
permitted lines, 144
phases of the Moon, 123-131
photography
  eclipses, 162-165
  photographic filters, 52
photosphere, 6

physicists, solar, 32
Physikalische-Meteorologisches Observatorium Davos. *See* PMO
pi (irrational number), 28
Pioneer missions, 11
plages, 21, 287
Planck curves. *See* black-body curves
planetary nebulae, 106-108
planetesimals, 65
planets
    bow shock, 10
    discovery around sunlike stars, 99-100
    exoplanets, 11
    laws of planetary motion, 45
    Neptune, discovery, 98
    planetesimals, 65
    Pluto, discovery by Tombaugh, 98-99
    positioning, Brahe, Tycho, 95
    Uranus, discovery by Herschel, 98
plasma
    gases, 66
    MHD (magnetohydrodynamics), 143
Pluto, 98-99
PMO (Physikalische-Meteorologisches Observatorium Davos), 286
Polaris (north star), 41
polarity, 20-21
position of the Sun, 40-48
prograde motion, 92
prominences, 23
    filaments, 23
    solar flares, 309
proton-proton chain, 68
    fueling the Sun, 34
protons (atoms), 66
Ptolemy, Claudius, 91-92

Q-R

quasars, 69
Queloz, Dedier, discovery of planets around sunlike stars, 100
quiet Sun, 5

R variable stars, 89
radar, 228
radiation
    black-body curve peak, Wien's Displacement Law, 86-87
    infrared, 22
radio waves, 22, 57
radioheliographs, Nobeyama Radio Observatory, 209-210
rainbows, 55-60
RAM (Reconnection And Microscale), 275-276
Ramaty High Energy Solar Spectroscopic Imager. *See* RHESSI
Ramaty, Reuven, 243
Rayleigh, Lord, 53
Reber, Grote, 228
Reconnection And Microscale (RAM) spacecraft, 275-276
red giants, 106
red sunsets, 53
retrograde motion, Copernican theory of the system of the universe, 92-93
RHESSI (Ramaty High Energy Solar Spectroscopic Imager), 243-244
Richard B. Dunn Solar Telescope, 197
Ring Nebula, 106
rockets
    data collected on eclipses, 239-240
    Saturn V rockets, 230
    V-2 rockets, 228
rods, 52

Roentgen, Wilhelm, 237
Roque de los Muchachos Observatory, 206
rotation of the Sun, differential rotation, 19-21
Rowland, Sherwood, 302
Russell, Henry Norris, 68, 103

S

Sacramento Peak
    Richard B. Dunn Solar Telescope, 197
    Sacramento Peak Observatory, 191
safety, looking directly at the Sun, 4
SAGE (Soviet-American Neutrino Experiment), 37
saros, 128-130
sarsens, 75
Saturn V rockets, 230
Scheiner, Christopher, detecting sunspots, 79
Schneider, Glenn, 178
SDO (Solar Dynamics Observatory), 273-274
seasons, 44-45
    eclipse, 126-127
    tilt of the Earth's axis, 44
SEC (Sun-Earth Connection), 313
SECCHI (Sun Earth Connection Coronal and Heliospheric Investigation), 138
seeing property of images, 188
seismology, 217
    helioseismology, 30-33
    seismic waves, 217
    seismometers, 217
shadows of the Moon, 127
siderostat telescopes, 196
SIM (Space Interferometry Mission), 100
Simon, George, 216
size, sunspots, 15
sky, Rayleigh scattering, 53

Skylab, Apollo Telescope Mount, 230-231
slow wind, 274
Small Explorers, flares, 243
SMM (Solar Maximum Mission), 232-233
measuring solar constant, 283
SNUs (solar neutrino units), 37
SOFIA (Stratospheric Observatory for Infrared Astronomy), 192
SOHO (Solar and Heliospheric Observatory), 8, 248-257
coronagraphs, 8-9
LASCO (Large Angle Spectroscopic Coronagraph), 9
Coronal Diagnostic Spectrometer (CDS), 254
Extreme-Ultraviolet Imaging Telescope, 30, 252-254
GOLF (Global Oscillations at Low Frequencies), 255
interferometry, 255-256
ions, 254
LASCO telescopes, 249-252
science, 249
solar wind, 257
UltraViolet Coronagraph Spectrometer, 254
VIRGO, 256, 286
Solar and Heliospheric Observatory. See SOHO
solar constant, 5
effect on climate, 290-291
measuring
Abbot, Charles Greeley, 282-283
Solar Maximum Mission, 283-284
spacecraft monitors, 286-287
Sun's energy, 282
sunspots, faculae, 287-290
solar corona. See corona
solar data, GONG+, 223
Solar Dynamics Observatory. See SDO

solar eclipses. See eclipses
solar energy
atmospheric warming, 294-295
greenhouse gases, 298-299
infrared radiation, 295-298
limiting human contribution to greenhouse effect, 299-301
equilibrium, 294
ozone hole, 301-304
solar flares, 22, 57
CMEs (coronal mass ejections), 310-311
Small Explorers, 243
space weather, 306-307
energy-storage mechanism, 307
prominences, 309
x-ray observations, 308-309
solar interior, wave oscillation, 220
Solar Maximum Mission. See SMM
solar neutrino units. See SNUs
Solar Optical Telescope. See SOT
Solar Optical Universal Polarimeter. See SOUP
Solar Orbiter, 276-277
solar physicists, 32
Solar Polar Imager, 275
Solar Probe, 274
Solar Radiation and Climate Experiment. See SORCE
solar rockets. See rockets
solar seismology, 30-33
solar spacecrafts, 229-235
solar surface
five-minute oscillation, images, 215-216
magnetic carpet, 266-267
wave oscillation, 218-220
solar system, measuring transits of Venus, 170-172
Solar Terrestrial Relations Observatory. See STEREO

solar wind, 9
SOHO, 257
Solar-A. See Yohkoh
solar-activity cycle, 21
faculae, 289-290
prominences, filaments, 23
space weather and solar flares, 307-311
CMEs (coronal mass ejections), 310-311
energy-storage mechanism, 307
prominences, 309
x-ray observations, 308-309
spectrum of light, 22
x-rays, solar flares, 22
Solar-B, 269-273
solid state of matter, 66
SOLIS (Synoptic Optical Long-term Investigations of the Sun), 202-203
solstices, 41-43
SORCE (Solar Radiation and Climate Experiment), measuring solar constant, 286-287
SOT (Solar Optical Telescope), 260
SOUP (Solar Optical Universal Polarimeter), 234
South Pole, 43
Soviet Union, Sputnik, 229
Soviet-American Neutrino Experiment. See SAGE
Space Interferometry Mission. See SIM
space shuttles, 233
Space Telescope Science Institute, 231
space weather, 305
aurorae, 311-313
solar flares, 306-307
CMEs (coronal mass ejections), 310-311
energy-storage mechanism, 307
prominences, 309
x-ray observations, 308-309

spacecrafts
Apollo, 229-231
monitors, measuring the solar
constant, 286-287
Nimbus series, 286
SORCE, 286-287
VIRGO, 286
RAM (Reconnection And
Microscale), 275-276
RHESSI (Ramaty High
Energy Solar Spectroscopic
Imager), 243-244
rockets. *See* rockets
SDO (Solar Dynamics
Observatory), 273-274
Small Explorers, flares, 243
SMM (Solar Maximum
Mission), 232-233
SOHO, 248-257
Solar Orbiter, 276-277
Solar Polar Imager, 275
Solar Probe, 274
Solar-B, 269-273
Spacelab, 233-234
Sun-synchronous orbit, 261
TRACE (Transition Region
and Explorer Spacecraft),
260-268
Ulysses, 234-235
Spacelab, 233
spectral lines, 57
absorption lines, exoplanets,
58-60
Balmer series, 69
converging series, 69
emission lines, 58
forbidden lines, coronal emis-
sion lines, 144-145
H-lines, 70
K-lines, 70
lines of ionized calcium, 70
permitted lines, 144
temperature of stars, 71
spectrographs, 139
velocities, 216
spectroscopes, 57, 139
spectroscopy, 138

spectrum, 22
elements in the Sun, 68
stars, 69
Balmer series of spectral
lines, 69
converging series of spectral lines, 69
H-lines, 70
K-lines, 70
lines of ionized calcium, 70
temperature, 71
wavelengths of light, 51
Sputnik, 229
stars
carbon cycle, 34
formation, 85
black-body curves, 85-87
H-H objects, 87-88
variable stars, 88-90
Milky Way, 63
spectrum, 69
Balmer series of spectral
lines, 69
converging series of spectral lines, 69
H-lines, 70
K-lines, 70
lines of ionized calcium, 70
temperature, 71
Sun as a star, 11-12
states of matter, 66
Stephenson, F. Richard, 134
STEREO (Solar Terrestrial
Relations Observatory), 138
stereo views, Solar-B, 271-273
Stonehenge as an observatory,
73-76
Stratospheric Observatory for
Infrared Astronomy. *See*
SOFIA
streamers, helmet, 142
subatomic particles, neutrinos,
35-38
Sudbury Neutrino Observatory,
38
summer solstice, 42
sun dagger, Chaco Canyon, 83

Sun dogs, 56
Sun–Earth Connection Coronal
and Heliospheric
Investigation. *See* SECCHI
Sun pillars, 56-57
Sun-Earth Connection Program,
Solar Polar Imager, 275
Sun-Earth Connection. *See* SEC
Sun-synchronous orbit, 261
Sunbeam (Yohkoh), 1993
Mercury transit, 178
sundials, gnomons, 47
sunlike stars, discovery of planets, 99-100
sunrises, 40-41
sunsets, 40-41
red color, 53
sunspots, 13
detection, 77-79
magnetic fields, 15-17
penumbra, 14
plotting latitude of spots,
Maunder's butterfly dia-
grams, 23
rotation, differential, 19-21
sizes, 15
solar constant, faculae,
287-290
solar-activity cycle, 21
sunspot cycle, 14-18
temperature, 15
umbra, 14
Sunspotter, 16
Super-Kamiokande experiment,
37
supernovae, 108
surface of the Sun, 4
Sweden
Institute for Solar Physics of
the Royal Swedish Academy
of Sciences Observatory,
206
Swedish Solar Telescope,
206-207
synoptic observations, SOLIS,
202-203

Synoptic Optical Long-term Investigations of the Sun. *See* SOLIS
systems of matter, angular momentum, 64-65
syzygy, 127

T

T Tauri stars, 89-90
telescopes
    Apollo, 230-231
    ATST (Advanced Technology Solar Telescope), 210-213
    Big Bear Solar Observatory, 201
    C. E. K. Mees Solar Observatory, 198-199
    discoveries of Galileo, 94
    filters, neutral density, 52
    GOES, 245
    GONG (Global Oscillation Network Group), 31, 222
    Hubble Space Telescope, Eagle Nebula, 65
    LASCO, SOHO, 249-252
    mountaintop, 190
    siderostats, 196
    SOHO, Extreme-ultraviolet Imaging Telescope, 252-254
    Solar-B, 271
    Sunspotter, 16
    Swedish Solar Telescope, 206-207
    towers, 189
    TRACE, 262
    Udaipur Solar Observatory, 208
    United States National Solar Observatory, 195-197
        McMath-Pierce Telescope, 196
        Richard B. Dunn Solar Telescope, 197
    vacuum telescopes, 207
    x-rays, 22, 239

temperature
    layers of the Sun, 8
    stars, spectral lines, 71
    sunspots, 15
Terrestrial Planet Finder, 100
tesla (T), 17
tilt of the Earth's axis, seasons, 44
time, mean solar time, 46
time zones, 47-48
Title, Dr. Alan, 178, 260, 266, 271
tokamaks (toroidal magnetic chambers with an axial magnetic field), 142-143
Tombaugh, Clyde, discovery of Pluto, 98
toroidal magnetic chambers with an axial magnetic field. *See* tokamaks
total eclipses, 6, 113
    African, 155
    Australian, 155
    Baily's beads, 117-118
    changing shape of the corona with the sunspot cycle, 120-121
    chromosphere, 7, 118
    corona, 7-9
    diamond-ring effect, 118
    European, 151-152
    future, 157-162
    observing every possible eclipse, 119
    phases of the Moon, 123-124
    silhouette of the Moon, 114
    videography, 165-167
    viewing and photography, 162-165
    visibility of the corona, 115-119
Total Solar Irradiance. *See* solar constant
tower telescopes, 189
TRACE (Transition Region and Coronal Explorer), 16, 260-268
    1999 Mercury transit, 178
    gas loops, 263-265

magnetic carpet, 266-267
magnetic reconnection, 265-267
moss, 267-268
telescope, 262
ultraviolet wavelengths, 262
trailing sunspots, 20-21
transition region, 259
Transition Region and Coronal Explorer. *See* TRACE
transits of Mercury, 177-179
transits of Venus, 169-180
transparency, 27
    observing the Sun, 188-189
tritium, 67
TSI (Total Solar Irradiance). *See* solar constant
turbulence, lake observatories, 190-191

U

U.S. National Academy of Sciences, 274
U.S. National Solar Observatory, Global Oscillation Network Group. *See* GONG
UARS (Upper Atmosphere Research Satellite), 284
Udaipur Solar Observatory, 208-209
UltraViolet Coronagraph Spectrometer (SOHO), 254
ultraviolet light, 22
ultraviolet wavelengths, 302
    Solar-B, x-rays, 271
    TRACE, 262
Ulysses, 234-235
umbra (sunspots), 14, 127
United Kingdom, Solar-B, 269-270
    stereo views, 271-273
    telescopes, 271
    ultraviolet wavelengths, 271
    views, 270-271
    x-ray views, 271

United States National Optical Astronomy Observatory, 196
United States National Solar Observatory, telescopes, 195-197
Upper Atmosphere Research Satellite. *See* UARS
Uranus, discovery by Herschel, 98

V

V-2 rockets, 228
vacuum telescopes, 207
Vacuum Tower Telescope. *See* Richard B. Dunn Solar Telescope
variable stars, formation of stars, 88-90
Variation of Irradiance and Gravity Oscillations. *See* VIRGO
velocities, 216
Venus, 169-180
vernal equinox, 41-43
videography, eclipses, 165-167
viewing eclipses, 162-165
    filters, 131
viewing the Sun
    coronagraphs, Mauna Loa Observatory, 199-200
    infrared observations, 191-192
    lake observatories, reducing turbulence, 190-191
    mountaintop telescopes, 190
    solar observatories versus nighttime, 185-189
    synoptic observations, SOLIS, 202-203
    telescopes. *See* telescopes
VIRGO (Variation of Irradiance and Gravity Oscillations), 286
    measuring solar constant, 286
    SOHO, 256

visible light, 22
volcanism, Io (Jupiter's moon), 218
von Braun, Wernher, 228
Voyager missions, 11

W

wavelengths
    light waves, 22, 51
    ultraviolet
        Solar-B, 271
        TRACE, 262
waves
    Earth, seismology, 217
    sun
        helioseismology, 217-220
        oscillation, 218-220
        p-modes, 218
        views of far side waves, 220-221
Weart, Spencer, 301
weather
    predictions, total eclipses, 148
    solar constant, 290-291
    space weather, 305
        aurorae, 311-313
        solar flares, 306-311
white dwarfs, 109
white light, 28
white polarity, 21
widths, eclipses, 132
Wien's Displacement Law, 86-87
Wilson, Richard, ACRIM, 284
wind, 255
    anisotropies, 255
    fast wind, 274
    slow wind, 274
    solar, SOHO, 257
Wolf sunspot number, 18
Wollaston, William, 57
woodhenge, 76
World Radiation Centre, 286
Wratten photographic filters, 52

X

X-class solar flares, 308
x-rays, 237-238
    coronal holes, 253-254
    grazing incidence, 239
    RHESSI, 243-244
    Small Explorers, 243
    solar flare classifications, 308-309
    Solar-B, 271
    telescopes, 22, 239
        GOES, 245
    Yohkoh, 241-243
XUV (extreme ultraviolet), 22

Y-Z

Yohkoh
    Solar-A, 241-243
    Sunbeam, 178

Zeeman, Pieter, Zeeman effect, 21
zenith, 41